FROM GALILEO
TO NEWTON

by
A. Rupert Hall

DOVER PUBLICATIONS, INC.
NEW YORK

Published in Canada by General Publishing Company, Ltd., 30
Lesmill Road, Don Mills, Toronto, Ontario.
Published in the United Kingdom by Constable and Company,
Ltd., 10 Orange Street, London WC2H 7 EG.

This Dover edition, first published in 1981, is an unabridged and
corrected republication of the work published in 1963 by Harper &
Row, N.Y., under the title *From Galileo to Newton 1630-1720,* as
Volume III in the series *The Rise of Modern Science* (original publisher:
Collins, London, 1963). A new Preface has been prepared by the
author especially for this edition. The original endpaper illustrations
have been omitted.

International Standard Book Number 0-486-24227-7
Library of Congress Catalog Card Number: 81-67038

Manufactured in the United States of America
Dover Publications, Inc.
180 Varick Street
New York, N.Y. 10014

For Alison

PREFACE TO THE DOVER EDITION

I have seized the opportunity in this reprint to correct a few errors made evident by work done since this book was first published. Otherwise I have left the original text unaltered because it has stood the test of time. If I were rewriting this book, I would alter a few emphases and smooth some harsh expressions, but I think it best that it should retain its original character and the influence of that great historian of science, Alexandre Koyré.

A. RUPERT HALL

GENERAL INTRODUCTION

The endeavour to understand events in nature is as old as civilisation. In each of its three great seminal areas—the Chinese, the Indian and the West Asian-European—men tried to find a logic in the mysterious and an order in the chaotic. They made many attempts, sometimes revealing strange similarities in these totally different societies, to express general truths from which particular events would follow as rational, comprehensible consequences. They tried to describe and analyse in order to understand, for men could not live in the world without seeking to assign causes to the things that happen in it.

One of these attempts to describe and analyse nature achieved remarkable fruition in Europe. Modern science is not merely European ; even before it had entered upon its triumphant age its establishment in North America and China had begun, and the origins of the intellectual tradition from which it sprang must be sought in Egypt and Western Asia. But the revolution in ideas which made modern scientific achievements possible occurred in Europe, and there alone, creating an intellectual instrument so universal and so powerful that it has by now entirely displaced the native scientific traditions of non-European societies.

The early stages of this Scientific Revolution began with the fifteenth century. The present book is concerned with its central, decisive stages. The Scientific Revolution was the effect of a unique series of innovations in scientific ideas and methods ; it gave the key to the understanding of the structure and relations of things. It was (and still remains) the greatest intellectual achievement of man since the first stirrings of abstract thought, in that it opened the whole physical universe—and ultimately human nature and behaviour—to cumulative exploration. Of its practical and moral implications we only now begin to have an inkling. For this colossal accomplishment Europe owed much to the Oriental world of which it then knew little. The vehicles of modern science, paper and printing, derived from China ; the language of science is still expressed in numerals devised in India ; Europe drew likewise on the East for its first knowledge of some phenomena (such as those of the magnetic compass), of some substances (such as saltpetre), and of some industrial techniques that relate to experimental science. But Europe did not borrow scientific ideas from the East, and in any case the borrowings had ceased before the rise of modern science began.

For this reason this volume will make only incidental allusion to science outside the Europeanised world. Europe took nothing from the East without which modern science could not have been created ; on the other hand, what it borrowed was valuable only because it was incorporated in the European intellectual tradition. And this, of course, was founded in Greece. The Greek philosophers, imposing no bounds on intelligence but those of the universe itself, set at the very root of the European tradition of science the ideal of an interlocking system of ideas sufficient to explain all the variety of nature. They were above all theoretical scientists, but at the same time they discussed critically the relationship between theories and the actual perception of events in nature. They began both observational biology and mathematical physics. Through most of two thousand years Europe continued to see nature through Greek eyes. Although the Scientific Revolution ultimately came as a reaction against the dogmatism inherent in the emulation of antiquity, it too drew its inspiration in part from neglected aspects of the Greek legacy. As Galileo admired Archimedes no less than Harvey did Aristotle, so the " mechanical philosophy " that flourished in the seventeenth century looked back to Epicuros and Lucretius. The Scientific Revolution did not reject Greek science ; it transformed it.

In a history of modern science it is unnecessary to describe the slow and devious process by which, after the fall of the Roman Empire, Greek science (with some accretions) was partially recovered and assimilated in Europe. The fresh exploration of Greek sources had a powerful effect on the fifteenth and sixteenth centuries, when medieval science seemed to have become sterile, although more justice must be done than the Renaissance allowed to the permanent merits of medieval scientific thought, especially in the study of motion. It possessed, as we can see, a certain richness which medieval philosophers themselves scarcely knew how to employ, but which gained its true expression in the hands of Galileo and his contemporaries.

How this came about is a principal theme of the present volume, *From Galileo to Newton*. Mathematics, for Galileo, was the essential key to the science of motion without which nature could not be properly understood : with its aid he showed that the Copernican system was plausible. Newton by the same means proved that system's truth. Further, this new outlook upon nature was enriched and confirmed by zealous attention to observation and experiment in astronomy, chemistry, natural history, physics and physiology. While the visible picture of the world was transformed, new mathematical theories reorganised some of these sciences and all were penetrated by the mechanistic concept. The resultant Newtonian universe was orderly, material, and " conformable to itself ". Small wonder, therefore, that science helped to usher in an age of reason from which superstition was to be banished, and in which enlightened men should enjoy all the fruits of their recently acquired wisdom and power.

A. RUPERT HALL

PREFATORY NOTE

This volume traces the changes in the spirit and ideas of science that took place between the publication in 1632 of Galileo's *Dialogues on the two Chief Systems of the World* and that of the final edition of Newton's *Mathematical Principles of Natural Philosophy* in 1726. It is not by any means a full record of the accomplishment of the crucial stage of the scientific revolution, but rather an attempt to characterise its nature. If it is largely concerned with the physical and mathematical sciences this is because the most profound changes of thought are found in them, and because they raised the major questions about rational understanding of the universe. At the same time I have tried to do justice, more briefly, to the intellectual content of the descriptive sciences which had not yet reached the same level of organisation, nor gained the same power of explanation, as the physical sciences. We should certainly not forget that the seventeenth-century mind was enlarged by a vastly wider, as well as deeper, acquaintance with Nature.

My great obligation to the many historians who have enlightened our knowledge of seventeenth-century science is partially expressed by indicating some of their work in the Bibliographical Notes at the end of the book. I have also thought it useful to record the sources of quotations, and to add a few supplementary details to the main narrative. The dates of birth and death of persons mentioned in it will be found in the Index.

Indiana University   A. RUPERT HALL

CONTENTS

CONTENTS

LIST OF ILLUSTRATIONS

LIST OF ILLUSTRATIONS

CHAPTER I

SCIENCE IN TRANSITION

1630–1650

What if the Sun
Be Center to the World, and other Starrs
By his attractive vertue and thir own
Incited, dance about him various rounds ?
Thir wandring course now high, now low, then hid,
Progressive, retrograde, or standing still,
In six thou seest, and what if sev'nth to these
The Planet Earth, so stedfast though she seem,
Insensibly three different Motions move ?
(John Milton, *Paradise Lost*, VIII, 122–130)

Among the young men admitted to the Mastership of Arts in the University of Cambridge in 1632 was a young scholar and poet who might but for various accidents have spent the rest of his life in academic quiet. About six years later, travelling in Italy, John Milton briefly met Galileo—in enforced retirement at Arcetri outside Florence as a result of the justification of the Copernican hypothesis that he had published in 1632. In 1665, before the Fire of London, Milton finished *Paradise Lost*, the splendid epic in which the ancient imagery of the Earth-centred universe spent its last creative force. He died nine years later, at a time when Isaac Newton was warmly engaged in defending his optical discoveries. During the years when Milton served republican England, he was also familiar with men who had plunged into science and the business of invention. He corresponded with Henry Oldenburg (1615?–77), who became Secretary of the Royal Society a decade later. He visited the house of Lady Ranelagh, beloved sister of Robert Boyle ;

he was known to one mathematician, John Pell, and probably to another, John Wallis ; he was acquainted with the ambitions of the reforming schemer, Samuel Hartlib, and may have met Comenius when he visited London at Hartlib's instigation in 1641. He lived through the transformation of science in England, though he was himself more concerned with the fleeting transitions of politics, and talked with some of the men who brought it about.

When Milton was born English science descended in almost limpid purity direct from the Middle Ages. When he died the Royal Society was in full course of building a new world, an earthly paradise perhaps though not a heavenly one. He was a man when Galileo was sentenced at Rome ; he lived through the whole active life of Descartes ; and having gone to school with Aristotle and Ptolemy he could have seen " at Mr Crosse's house in Oxford " the very beginning of the long road that led to Rutherford. Caught between the past and the future Milton's present held the fall of classicism and the rise of modernism, the reluctant yielding of Puritanism before deism, the passage of the new science from diffidence to mastery.

The first thirty years of the seventeenth century had shaken the old order of things but by no means disrupted it. Traditional science so far revealed astonishing resilience and the new had not yet acquired an outlook positive enough to take its place. Schools and universities all over Europe continued to teach the comfortable doctrines of natural philosophy and medicine drawn from classical authors much as they had done for two centuries before. To an ordinary observer of the learned world in Milton's youth only two groups presented themselves as markedly dissident. The more serious consisted of those astronomers who persisted in upholding the belief of Copernicus—still after some eighty years regarded by all but a few enthusiasts as fantastically absurd—that the Earth and planets circle a stationary sun. So feeble had the arguments in favour of Copernicus seemed and so evident the fixity of the Earth that it was only in 1616 that Copernicanism had been condemned by the Catholic Church, save as a calculating device. The real battle between traditional and revolutionary ideas in astronomy had been long delayed, and when it came its violence was largely confined to Italy. Elsewhere the transition from scepticism to acceptance of the heliostatic system occurred peacefully enough in the second quarter of the century ; but before 1620 there were few Copernicans anywhere. In France, for instance, Marin Mersenne (1588–1648)—later to become a central figure of the scientific movement in his country—published in 1623 a work in which he showed, very fairly, the weakness of the Copernican hypothesis. He did not change his mind until about 1630. Descartes (1596–1650) had probably swung over rather earlier, yet he always hesitated to avow himself openly a Copernican. Learned opinion in France was broadly of Mersenne's mind.

In England William Gilbert (1540–1603), physicist and physician, had made the rotation of the Earth the pillar of his magnetical philosophy without following Copernicus in setting the Earth free to revolve about the sun. There were others, however, who followed the sixteenth-century example of Thomas Digges in taking the opposite view, among them the Gresham College Professors Briggs and Gellibrand. And Sir Henry Savile, in founding a chair of astronomy at Oxford in 1619, had wisely stipulated that the system of Copernicus should be taught alongside that of Ptolemy. In fact, though few Englishmen as yet subscribed firmly to the new celestial system, many of the well informed recognised the imperfection of the Ptolemaic, and looked for some kind of compromise such as that offered by Tycho Brahe (1546–1601). For Tycho made the five planets spin around the Sun, while the Sun and Moon revolved about the Earth ; hence the fixity of the Earth was maintained although the relative motions were the same as in the Copernican system.

It did not follow that—outside Italy—adherence to Copernican ideas was regarded as reprehensible. Moreover, some scholars though sceptical nevertheless made use of Copernican tables and astronomical constants, as Erasmus Reinhold (1511–53) had done years before in compiling his *Prutenic Tables*. In many places discussion of the rival theories took place without sharpness, and there was no open crisis even in Italy before 1632, despite the cardinals' decision of 1616. The career of the greatest of early seventeenth-century astronomers, Johann Kepler (1571–1630), was not affected by his unconcealed attachment to Copernicus's system. The storms in Kepler's life were not occasioned by his scientific opinions, though indeed when he died two years before Galileo's trial it might have seemed that he had lived in vain. In the strategy of science Kepler's discoveries are among the greatest, and tactically they yielded the most solid support for the heliostatic view that the age could furnish. But no echo of Kepler's laws of planetary motion is perceptible until a decade after his death, while in his lifetime he was best known for fantastical and absurd speculations— and for his optics. Even Galileo (1564–1642), besides failing to elucidate the significance of Kepler's discoveries (in public at any rate), seems to have had little wish to link his own rational defence of Copernicanism with the supposed whims and fancies of the Imperial Astronomer. The Pythagorean mysticism, the far-fetched ratios and musical harmonies of Kepler's books repelled many who sought, rather, one single solid reason for supposing the Earth to move.

Galileo's history is very different. Like Kepler a fairly early convert to the new astronomy, in 1597 he confessed his fear of declaring himself lest he should be mocked. Throughout his career he taught his pupils the Ptolemaic system and it is probable that he never lectured publicly on the physical truth of the Copernican. Certainly he denied that he had ever done so. However, he did discuss the old and the new astronomy in private before 1632 (as it was lawful for him to do) and among his pupils he found some notable converts for Copernicus, such as Benedetto Castelli (1577–1644) and Bonaventura Cavalieri (1598–1647). From 1610 onwards he wrote plainly in favour of Copernicus and against any attempt to suppress preference for the new astronomy, or discussion of its tenets. Galileo had become famous throughout Europe as the first to turn the telescope to the heavens, as the discoverer of Jupiter's satellites and the mountains of the moon, of the spots on the sun and the phases of Venus, so that it might seem, with his authority as an investigator reinforced by his vigour as a polemical writer, that Galileo's opinion would have carried great weight in favour of Copernicus even before he published the *Dialogues on the Two Chief Systems of the World* (1632). This would be too simple a view. Like Kepler, Galileo had won few converts before 1630, most of them among his circle of friends and pupils. His discoveries and writings did two things. They provoked the first really powerful counter-attacks against the new doctrines in astronomy, and they also multiplied the number of these new doctrines. The question was no longer simply whether the mathematical system of Copernicus was physically correct or not. For a time at least the situation, the decision for or against traditional ideas, was not clarified but rather confused by the new discoveries made by Galileo and others.

Criticism of Galileo took three forms. First, there were attacks on the truth and originality of his observations—the former more understandable because it proved very difficult in the early years to repeat them, until Galileo had distributed a number of his own telescopes which were much superior to those bought in the opticians' shops. Secondly, his interpretation of what he saw—that the moon is rugged and mountainous, that the Earth reflects light like the moon, that the sun has dark blemishes whose movement demonstrates its rotation and so forth—was doubted by many who admitted the ocular evidence. And thirdly it was not allowed by his opponents that the new sight of the heavens given by the telescope in any way confirmed the Copernican pattern of celestial motion. Copernicus' innovations in astronomy had been essentially geometrical ; Galileo's were essentially physical. It was possible to tie the two together—though Galileo only attempted to do so in detail in the *Dialogues* of 1632—but it was equally possible to avoid doing so.

Galileo's critics could quite reasonably hold that the new discoveries did not prove the truth of the Copernican system though they might (and did) destroy the Ptolemaic.

They could take this position by following the example of the great Danish astronomer Tycho Brahe, who had rejected both Ptolemy and Copernicus. Tycho's own system of celestial motions had the merit of being theoretically equivalent to the Copernican, without the apparent defect of ascribing motion to the Earth ; it made possible a scientifically adequate geostatic astronomy, irrefutable by any test of observation that Galileo or anyone else could impose upon it. As such it was adopted by many writers, especially by orthodox Catholic astronomers such as Giambaptista Riccioli (1598–1671). The Tychonic system was effectively current long after the Ptolemaic was defunct, surviving until after mid-century. Relying on this modern geostatic conception anti-Copernican and anti-Galilean astronomers like the Jesuit Christopher Scheiner (1575–1650) could not only accept Galileo's physical observations of the new celestial phenomena, but claim them for themselves. In the same fashion Tycho, though anti-Copernican, had argued that there were no celestial spheres and that comets were true celestial bodies. Acceptance of the reality of Jupiter's satellites and of sunspots put the critic in a far stronger and more flexible position than that which had been adopted by Galileo's early traditionalist opponents, who had simply decried everything seen through the telescope. It could now be argued that Aristotle was in error only in so far as he had unfortunately lacked such a device for exploring the sky. Well and good : mountains on the moon prove it is not a perfectly crystalline sphere, but they do not prove that the Earth moves.

In the years just before the publication of Galileo's *Dialogues* there was little reason to anticipate a violent revolutic.1 in astronomical theory. Fresh information had come in swiftly since the first use of the telescope in 1609 but it seemed that its import could be neutralised by accommodating it within the old framework. The spread of Copernican ideas was slow and undramatic. They were still opposed by most learned men and by virtually all the mathematical astronomers except Kepler. The latter's accurate solution of the problems of planetary motion was universally ignored. The innovators themselves were not completely agreed on the new shape of the heavens ; Galileo was conservative in denying that comets were heavenly bodies, Kepler in denying that the universe could conceivably be infinite. On lesser matters—the size of the heliocentric orbits, the strange appearance of Saturn, the cause of terrestrial tides—confusion reigned among them. Yet, within about a quarter of a century, the issue was decided in favour of Copernicus and the Earth was henceforward as likely to be considered flat as fixed. The decrees of 1633 were issued at the very moment when they were useless.

The other dissident group that reveals some coherence in the early seventeenth century was in the long run of far less significance in the development of science, and was (perhaps naturally) proportionately more noisy in its own time. The iatrochemists (chemical physicians) distorted a good case against traditional medicine, whereas the astronomers were in the right even though they could not prove it. Just as the latter attacked the authority of Aristotle and Ptolemy, so their companion innovators attacked that of Galen and the whole long line of Graeco-Arab physicians descending from him. In place of Copernicus they had his near-contemporary Paracelsus (1493–1541) ; for the heliocentric system the therapy of chemically-prepared medicaments; for the mystique of numerical relations the mystique of fire as the sovereign of chemical and bodily action ; for the decrees of the Church the condemnations of the established faculties of medicine. And as the war against new ideas in astronomy was hottest in Italy, so the war against them in medicine was hottest in France. Elsewhere—in Germany, the Low Countries, England—Paracelsan ideas (or alternative more rational versions of them) were allowed to make slow headway, just as Copernican notions did.

There was never so marked a change in opinions about the proper kind of remedies to use against disease as that which took place in astronomy. The older herbal medicines—" Galenicals "—continued to hold their place in the pharmacopoeias, if always in retreat. While only a few chemical preparations were admitted into the first edition of the *London Pharmacopoeia* in 1618, their number increased steadily with each reissue during the seventeenth century. Approaches to medicine, physiology and chemistry proper that owed something to the teaching of Paracelsus reached their maximum influence about mid-century ; thereafter the effect of Paracelsus declined again. Elements of mysticism were gradually pruned away till only a rational basis remained, just as happened with the celestial harmonies of Kepler. There was an increasing tendency for alchemy, like astrology the disreputable companion of astronomy, to be set aside as an aberrant variation of true chemical science. For the first matter-of-fact manuals of empirical chemistry had appeared in the first years of the seventeenth century, and their pattern was developed with further elaboration.

The comparison between Paracelsan chemico-medicine and the new astronomy indicates that the struggle between tradition and innovation in science was not necessarily (or simply) one between wrong and right as judged by later standards. The iatrochemists were no less sure of their innovations than were the Copernicans. They argued as tenaciously and more volubly, they were no less ready with experiential proof and philosophical reasoning to justify their case. With no less justice they could resent the dead weight of tradition that opposed them and the intolerance of authorities ; they could appeal with no less effect to the virtue of the open, inquiring mind and of the experimental method. If, in the end, their views have been found to hold but a dim perception of the truth it can equally be said of the Copernicans that they had seized upon but the first clue to modern astronomy. Just as certain pages of Kepler, contrasted with the plausible sanity of some anti-Copernican astronomers, cause one to wonder which was the side of the angels, so the Paracelsan insurgence underlines the current of fantasy in the ebullience of seventeenth-century science.

For despite the error of its content and the weakness of its methods, the legacy of ancient science was eminently rational and logical; such it had been in the beginning among the Greeks and as such it was remoulded by the scholastic philosophers of the Middle Ages. The true scientific tradition had invariably opposed the magical view of nature, the view that events are governed by spirits or demons or other unknowable forces not obeying the normal laws of cause and effect. Such a view was always present, it was at the root of popular superstitions and of beliefs that learned men had transported out of superstition into science at various times. But the conscious effort of the learned was always in the opposite sense. The distinction between sanity and superstition was not always easy to draw : at one time the notion that the moon causes the ebb and flow of the sea—based on the connection between lunar phase and tidal flow well known to sailors—was regarded as a superstition like that of farmers who would only plant their seed at new moon, or of herbalists gathering plants at full. How could the remote moon push and pull the water of the sea ?

The more full of strange marvels the world was found to be, the more surprising the discoveries made in astronomy, chemistry, zoology and botany, the less possible it seemed to say what can be, and what cannot be. Nature was so far more rich than ever reason had supposed it. That a woman in the Rhineland should give birth to a hundred rabbits or that emeralds should grow like grass in the mines of Java was hardly a stranger tale than that of the sensitive plant, a cyclopian calf, Saturn's ring, or the animalcules of rain-water. Wandering among wider intellectual horizons with traditional guides falsified or disturbed, it was often difficult to separate unconscious self-deception from deceit, and the irrelevant from the crucial. Some men, like Galileo, who rarely even in his letters spent time on what was trivial or absurd, had an instinct for the significant ; others, like Kepler and even Newton on occasion, were less sure in their touch. In the huge bulk of seventeenth-century scientific writing, besides a great deal of triviality, there is much produced by fantasy, from Kenelm Digby's weapon-salve (applied not to the wounded man but to the weapon that struck him), and Paracelsus' *archeus* (an intestinal chemist) to van Helmont's *alkahest* (the dissolvent of all things). Ordinary medical practice was replete with revolting absurdity. Not that all the extravagances of seventeenth-century science were superstitious in origin; some—like the theory that thunder is caused by an explosion of celestial gunpowder—merely invoked a rational effect in a mistaken fashion. But it is not difficult, even though it is often not very enlightening, to prove that science was still penetrated by the magical view of nature ; belief in witchcraft, at least, was practically universal.

Iatrochemistry was born of superstition, for Paracelsus' view of nature was deeply imbued with magic even though he gave it an empirical dress. Few of his seventeenth-century followers shared his belief in the possibility of reconstituting a living bird by art from its burnt ashes and similar fantasies, but the traces of such belief were still with them. Similarly the magical view is stamped upon astrology—now almost utterly discredited among serious astronomers—and on the general literature of alchemy (which was in fully rational terms accepted by practical chemists). Its lingering influences on medicine and the lore of animals and plants were still plainly discernible. Respectable naturalists continued to credit the spontaneous generation of frogs and insects into the second half of the century, which is once more the decisive epoch in this respect. By its close there was little left of the magical outlook, of Paracelsism and esoteric science ; the Pythagoreanism of the Renaissance, its Faustean spirit and natural magic, had all quite gone—not without some benefits to rational science on the way.

The route to complete rationalism in science was hard to follow. The breakdown of the traditional academic certainties of the Middle Ages, combined with the ambition to find fresh truths in field and wood and mine, in the simple knowledge of ordinary men, by deserting intellectual sophistication for the plain ground of experience and commonsense, could yield strange results. Among them, the fact that Joseph Glanvill (1636–80), one of the most eloquent champions of the young Royal Society, and Cotton Mather (1663–1728), one of the most influential exponents of science in the American colonies, were also in their respective countries the most deluded enemies of witchcraft. The emphasis on empiricism and plainness could promote naivety, leading away from the intellectual exploration that is the true course of science ; conversely, recognition that natural phenomena are more complex than scholastic philosophy allowed could end by making the world an unfathomable mystery. Science had lost much of its established intellectual discipline in the sixteenth century, and only acquired a new one in the second half of the seventeenth. Crudely considered, the experimental method tended towards indiscipline with its suggestion : anything is possible, therefore let it be tried. The Royal Society did not scruple to put a girdle of flame about a scorpion. For those who came to scoff there was a ready source of ridicule in the monstrous labours of science that brought forth very small mice, weighing the air—or, as Shadwell had it in *The Virtuoso*, practising the theory of swimming on dry land. By a strange reversal the commonsense of Stubbe or Shadwell or Swift could become the torment of philosophers who had only a little before used their own appeal to commonsense to overturn scholastic science.

Together with the vast expansion of scientific observation and experiment in the seventeenth century, and a still more important breach in the barriers limiting scientific ideas and theories, there was a great enlargement of the living company of scientists. There have always been popular expositors of science from Pliny onwards but in the main the authentic activity of the later Middle Ages had been the work of a small group of professional academics—and there has never been a tougher intellectual discipline than that of the fourteenth-century philosopher, however little relation his ideas

27

bore to the realities of nature. Basically the same is true of the seventeenth century, for the scientific revolution was effectively the achievement of academic professionals like Galileo, Kepler, Cavalieri, Wallis, Newton, Hooke, Leibniz and Huygens—to name a few at random—and of non-academics who were no less deeply committed to science like Harvey, Fermat, Descartes, or Hevelius. It would be stupidly pedantic to try to make a distinction between these groups. On the other hand there was a large crowd of dabblers in science, most conspicuous in England perhaps, but found elsewhere too, ranging from grandees like Prince Rupert and Leopold de' Medici to humble gaugers, surveyors, country clergy and physicians. Some few made contributions of real merit ; others, John Evelyn for instance, enjoyed a reputation quite disproportionate to the value of their scientific attainments.

The *virtuosi* or *curiosi* whose names appear so frequently in the papers of Mersenne and Oldenburg were typical of the seventeenth-century scientific scene, particularly after the first few decades. In a sense they descended from the Renaissance virtuosi who collected manuscripts, antiquities, medals and sculpture and it is not surprising that they first appear in Italy. The friends of Galileo who are immortalised in his dialogues, Salviati and Sagredo, were men of this kind just as Prince Federigo Cesi, founder of the Academia dei Lincei, to which Galileo proudly belonged, was the prototype of seventeenth-century patrons of science. Later the virtuosi become more visible to history in the massive correspondence of Fabry de Peiresc (1580–1637) and Marin Mersenne in France and of Samuel Hartlib and Henry Oldenburg in England, followed in turn by periodicals such as the *Philosophical Transactions* and the *Journal des Scavans* which were supported by the virtuosi and printed much of their writing. They formed the major section of the public interested in scientific and medical matters ; for the most part they were educated, they were often influential in their professions, and sometimes they possessed power at Courts. If the importance of the contributions of individual virtuosi to seventeenth-century science is slight, that of the total of their writings is not, and the results of their interest in others more gifted than themselves were often creditable. Sir Jonas Moore (1617–79) is more likely to be remembered as the patron of the first Astronomer Royal, John Flamsteed (1676–1719), than for his own textbook on mathematics. Generally, however, the interest of the virtuosi in science did not favour its sterner branches, a fact not without effect on the scientific movement. They turned easily to natural history, gardening and the cultivation of rare plants ; to the curiosities of nature and medical practice (petrifying springs, mineral waters, strange geological formations, monstrous births and autopsies) ; or to the applications of science in painting, architecture, music and war. Sometimes—with the greatest of all scientific patrons, Louis XIV—their interest was exclusively utilitarian, in submarines, ballistics, fortification, waterworks ; a few joined the " projectors " of canals, mines and new industrial works. Others turned to chemistry and hazardously dosed their neighbours. Much of the rich diversity of seventeenth-century scientific writing stems from this source and it is not surprising to find in it the same mixture of perceived truth and unconscious fallacy that occurs in the *Vulgar Errors* (1646) of Sir Thomas Browne—the greatest of all the English virtuosi.

To these men in whom a love of natural science sometimes mingled with connoisseurship in art, a taste for history, or a desire to explore remote parts of the globe, must be ascribed much of the breadth of the scientific movement in the second half of the seventeenth century. The tendency to broaden science—particularly to treat seriously such subjects as metallurgy, geology, botany and zoology, and the near-medical sciences like comparative anatomy and embryology, all of which had virtually no place in the medieval tradition despite their Greek antecedents—began far back in the Renaissance. It was not created by the virtuosi, but they gave it greater depth and extended it further. On the other side of the balance, however, they were responsible for much of the extravagance of seventeenth-century science, partly no doubt because they were not trained in mathematics or medicine, partly because they were not always men with strong minds and a firm grasp of reality. If they were not the only readers of Robert Fludd, Athanasius Kircher, Kenelm Digby, van Helmont and others of the more esoteric authors of the time—who, even though their importance is considerable, lie off the mainstream of rational scientific development —they were the readers most influenced by such writers. The virtuoso spirit was a good servant of rational science but a poor master. It could promote the accumulation of knowledge without giving it the systematic, theoretically organised character that belongs to science.

Ultimately the professional intellectual attitude had to dominate, even at the cost of some narrowing of the scientific attack. The struggle with amateurishness left its mark on institutional history, on the Royal Society, the Accademia del Cimento and the Académie des Sciences. And it contributed in the eighteenth century to the growing isolation of men of science from scholars. But for the moment, in the second quarter of the seventeenth century, the spread of a virtuoso interest in science was a source of strength and a provision of opportunity.

Besides the virtuosi, there was another group in the scientifically preoccupied public whose contribution to the scientific movement was becoming steadily more noteworthy. Before modern times there are only dim images of the highly technical crafts—those of the incipient engineers (surveyors, millwrights, military engineers, smiths and clockmakers), industrial chemists (metal-smelters, assayers, distillers, pharmacists) and instrument-makers (opticians, rule-makers, gaugers). Certainly in the sixteenth century many of these craftsmen were educated and alert ; they wrote books and made new inventions in their trades. The business of navigating ships in particular became highly mathematical and some serviceable manuals on this difficult art were compiled by practical men like the Englishman William Bourne (fl. 1565–88). While it is true that in the seventeenth century, as before, few of these educated craftsmen ventured to write on subjects remote from their trade the greater volume of such technical writing is remarkable. Some of it appeared in scientific journals like the *Philosophical Transactions* and technical work was taken seriously by the professed scientists. A few craftsmen had real theoretical capacity : Michael Dary, a gauger of wine-casks by trade, corresponded with Isaac Newton on mathematical topics. Naturally the craft closest to science was instrument-making, which became highly diversified. Instrument-makers and scientists co-operated in the improvement of optical instruments and Robert Hooke (1635–1703) records many visits to the celebrated clock-maker Thomas Tompion ; as

> [Saturday 2nd May 1674]. Told him the way of making an engine for finishing wheels, and a way how to make a dividing plate.; about the forme of an arch ; about another way of Teeth work ; about pocket watches and many other things.[1]

How much the excellence of Tompion's clocks may owe to Hooke's conversations is not clear. Some instrument-makers made for sale the recent inventions of scientists, such as Napier's " bones ", Gunter's scale, and William Oughtred's spiral calculator, the first logarithmic rule. The optician Reeves was pressed by the Royal Society to develop Newton's invention of the reflecting telescope. Others from their experience and insight improved on scientific inventions ; thus the clockmaker William Clement devised the first really practicable escapement for the pendulum clock invented by Huygens. The evolution of the marine chronometer involved constant interaction between science and craft. Other relatively new scientific instruments such as the microscope, telescope, barometer and thermometer were normal articles of trade long before the end of the century.

The transition of science to instrumentation was very rapid in the first half of the seventeenth century—indeed it is easy to over-emphasise its significance. The first great burst of discovery with the telescope of 1609–12 was the first and last effort of the " Galilean " device. Its usefulness was soon exhausted. All later refractors of more than opera-glass power have employed the biconvex lens combination first described by Kepler in 1611, and introduced into astronomy about 1640. Within a few years magnification was increased vastly. The microscope was not used for serious scientific observation before 1625 nor were the first notable studies made before the 1650s. The first barometer for estimating the pressure of the atmosphere was made in 1643 ; attempts to procure experimental vacua by pumps followed a decade later. The thermometer was scarcely used systematically at all, though it was the first of the new instruments of science. With the exception of Tycho Brahe's work in astronomy higher accuracy of measurement was not regarded as a major objective during the early stages of the scientific revolution so that it was only in the second half of the seventeenth century that a number of fresh devices were introduced—the pendulum clock, telescopic sights, the bubble-level and screw-micrometer—to transform the notion of scientific exactitude. The standards prevailing about the time of Newton's death (1727) were utterly different from those of the age of Galileo and Kepler. Although the drive behind this change came from the scientists it was largely made possible by the success of the instrument-makers who served their needs.

The desirability of precision in measurement in such sciences of direct observation as astronomy had presented itself readily enough, even if it was difficult to discover means to attain accuracy. The adoption of quantitative procedures in experimental science was a more complex process. In many kinds of experiment the very notion of measurement was a distinct concept that had to be definitely formulated, and once it was formulated became by that fact an integral element in the whole experimental procedure. The principle of determining the position of a star in the sky is the same whether the measurement is crude or exact, but quantitative procedures completely alter, or even constitute, the nature of some experiments. To attempt to measure adds a whole new dimension

The frontispiece to Galileo's *Dialogue* of 1632: Aristotle, Ptolemy and
Copernicus debate the structure of the universe

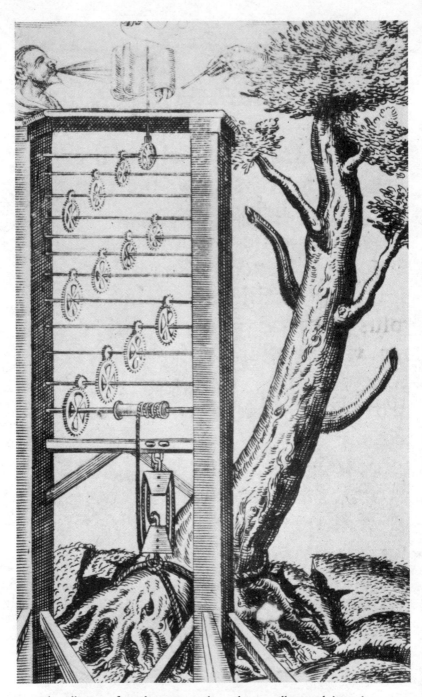

The alliance of mathematics and mechanics, illustrated by John
Wilkins in *Mathematical Magick*, 1648

to experimental science, no less than to the body of theory relating to the experiments. It is rather missing the point, therefore, to refer to the "rise of experimental science" in the seventeenth century as though one kind of experiment was exactly equivalent to another. It is not at all the same thing to try the effect of planting cider-orchards in Norfolk as to determine by pendulum experiments the acceleration due to gravity. Experiments of the "Let's see what happens" type were common enough in seventeenth-century science (especially in chemistry and medicine where the possibility of prediction was poor), and it is certainly a measure of the new scientific enterprise that they were made. Much of value was learnt from them, not least the desirability of performing such experiments in a more satisfactorily quantitative manner. But this—as the critics of Francis Bacon have always pointed out, perhaps with some injustice to Bacon himself—was far from being all that the development of experimental science meant. To be designed at all an experiment involving measurement necessitates a theoretical pattern ; it can never be a random product. If the experimenter selects for measurement one or more events it is because the significance of these events has been foreseen, and indeed it is often crucial to decide in advance which of the events in the experiment are to be measured.

Once more, the early seventeenth century was a phase of transition in the history of scientific experimentation. The experimental method of inquiry received the powerful advocacy of Bacon, and was exemplified in the work of the medical chemists, of Gilbert and his predecessors in the study of magnetism, and of others like Cornelius Drebbel (1572–1634) who hovered rather dubiously on the frontiers between science, technology and natural magic. Such experimental science, far from systematic, was still directed to exploiting nature's wonders as much as to revealing nature's laws. It belonged at least as much to the past as to the future. Far more decisively novel—though indeed it had been anticipated—was the incipient use of experiment as a method of proof, whereby the result served to verify or falsify a previous expectation. This role of experiment was appreciated by Bacon (though he had not emphasised it), was illustrated in Harvey's *De motu cordis* (1628), but was above all inculcated in the major works of Galileo. In all these, however, the element of precision is lacking. Harvey's experiments were little more than demonstrations of, for example, the action of the valves in the veins of the arms of a living subject, and his famous quantitative argument on the flow of blood through the heart rests on an approximation, not a measurement. Many of Galileo's experiments (or rather, appeals to experience) were rhetorical ; they were not reports of events made to occur in a precise fashion. This is not to deny that Galileo made experiments on floating bodies, thermoscopes, pendulums and many other things, including experiments on falling bodies and inclined planes (as we have discovered recently through the work of Stillman Drake), which involved exact measurements. But generally speaking, only in the second half of the century, mainly in the work of men born when Galileo was already aged, did experiment become a meticulous tool of science.

Even as late as 1630, in fact, systematic observation had proved a far more constructive method in science than had experiment. It was by observation that solid knowledge of human anatomy had been built up since the early sixteenth century ; that Tycho had provided the materials for far exceeding the revolution of Copernicus ; that botany and zoology had assumed organised form. Observation had brought cosmological theory to the point of crisis. However much the establishment of a new science was indebted to the experimental inquiry later, as yet—in Tycho, in Harvey, even in Galileo—mere observation of what is happening all the time was almost enough to destroy the formal rigidity of the old. To have the use of one's eyes—if one knew how to direct them—was to see that what Aristotle or Ptolemy had described was false and to find reason for a fresh idea of nature. And it was in this idea that the explosive force of the scientific revolution lay.

Notwithstanding the excitement and passion that ensued after the invention of the telescope, the force of scientific ideas scarcely began to act during the first quarter of the seventeenth century. They were still locked up in unwritten books and obstructed by the conservative scepticism which had so long delayed the acceptance of Copernicanism. The ideas on the motion of bodies already set by Galileo before his ablest pupils, or discussed between Isaac Beeckman (1588–1637) and Descartes at the other end of Europe, were still unknown to science at large. The mechanical philosophy had not passed beyond emulation of Epicuros. Geometry was still held in the mould of Apollonios and Pappos, algebra—despite enormous progress—had hardly disentangled its first principles. Chemistry was a chaos of technical recipes and speculative philosophies of nature. Nothing in the science of these early years is clearly detectable as modern, or even as pointing unambiguously to the revolution in ideas and methods that was so soon to happen. No one could have perceived that the science of antiquity, so long the model and the source of inspiration, had made its last contributions to the formation of modern European science, which was to be grounded not on tradition but on its own observational and experimental processes, its own metaphysic, and its own philosophy of knowledge.

CHAPTER II

THE GALILEAN REVOLUTION
IN PHYSICS

Now these things take place in motion which is not natural, and in materials with which we can experiment also in a state of rest or moving in the opposite direction, yet we can discover no difference in the appearances, and it seems that our senses are deceived. Then what can we expect to detect as to the Earth which, whether in motion or at rest has always been in the same state ? And when is it that we are supposed to test by experiment whether there is any difference to be discovered among these events of local motion in their different states of motion and of rest, if the Earth remains forever in one or the other of these two states ?
(Galileo, *Dialogues on the Two Chief Systems of the World*, 1632)

The transcendent discovery of the seventeenth century in explaining the universe to itself and posterity was the universal power of motion. Throughout the ages the problem faced by natural philosophers had been, in broadest terms, the duality of identity and change. The whole universe, the materials composing it, the events that occur in it, are in one sense always the same, in another always different. Everything that happens is unique, yet it is part of a pattern; and the substance that entered into the singular drama of one human life returns to the common store of matter :

Why may not imagination trace the noble dust of Alexander, till he find it stopping a bung-hole ?[1]

Ex nihilo nihil fit. For some of the ancients even time was not unique since in the cycle of the years the past would be restored and events recur; and no one believed that the substance of material things, in this world of transience, was ever ultimately created or destroyed. Many attempts have been made to account for the appearance of change, of growth and decay ; that of the seventeenth century, deriving from the atomism of Democritos, Epicuros and Lucretius, assumed the existence of material entities whose identities were fixed and unalterable, participating in motions both various and variable. The sameness of the universe reflected the constancy of matter, its changes reflected the variability of motion. And this mechanistic philosophy was applied not merely to obvious phenomena of change (where rearrangement of constituent parts occurs) but to phenomena where alteration does not superficially appear (as when motion occurs without rearrangement of the parts). For it was believed that nothing is truly at rest ; besides many " languid and unheeded motions "[2] there were those only revealed by phenomena in which, at first sight, movement played no part at all.

Thus the seventeenth century reversed the view of Aristotle for whom motion had been a kind of change—change of place. Instead change—and indeed all scientific phenomena—became manifestations of motion. Thereby the science of motion was made the primary science, since in the last resort the explanation of all things happening in nature should end in a description of the motions of the changeless material entities involved in them. There was nothing new in the basis of this idea. Aristotle's *Physics* had itself been devoted to a discussion of the nature of motion and change while fourteenth-century philosophers had returned to the same problems, formulated in more mathematical terms. What was new was the concept of the universality of motion ; not only that motion occurred everywhere, but that it was the same kind of motion everywhere. Motion, it now seemed, was invariably subject to the same laws, from the scale of the sun through the scale of an apple to the scale of the least particles of matter.

Logically, then, the application of the science of motion to celestial bodies at one extreme, or to the fundamental particles of matter equally far removed from the ordinary scale of things at the other, could be made by extension from what could be discovered of the science of motion within the range of immediate experience. For Newton, indeed, this was the necessary " foundation of all philosophy ".[3] Moreover as the science of motion developed it became clear that its starting-point, the Laws of Motion from which everything else could be derived with the aid of suitable definitions according to mathematical strictness, did not have to rest solely on the dubious validity of such experimental tests as could be devised by men upon the surface of the Earth. It seemed that the Laws of Motion *could not be false*, that is, it was inconceivable that any alternative propositions could be valid. This was tantamount to saying, as we do when we allege that two bodies cannot be in the same place at the same time, that intelligence revolts at the thought of the falsity of the laws of motion. It was superfluous to attempt to test the laws by experiments under conditions that in any case could never correspond to those under which the laws would be rigorously true. Such experiments did not prove the laws but illustrated their applicability.[4] Accordingly, the science of motion did not depend altogether on inference and the accuracy of experiments • for its application to those bodies beyond the reach of experiment, unless it should be true that Nature was not everywhere uniform and " conformable to herself"—which no seventeenth-century philosopher or scientist was prepared to suppose. This was the truth that Newton's *Principia* was to demonstrate near the end of the century.

Nor is this all, as the *Principia* also shows. For if the phenomena of nature arise from the motion of particles, the science of motion provides the essential key to physics and beyond that to chemistry and even physiology—since these sciences depend on the physical properties of matter. Thus the mechanistic explanation of things could develop as an endless chain, beginning with the laws of motion and stretching further and further into successively more complex departments of science ; always however traceable back to firm roots in the laws of motion that could not be false. Here, potentially at least, lay the unshakable foundation for all scientific knowledge, independent of the uncertainties of deduction or induction, experiment or observation.

This path for the development of scientific thought was by no means an open one at the start of the century, when it was obstructed by quasi-occult leanings towards the magical view of nature which, inevitably, denied the validity of the endeavour to fragment the natural world into an assembly of mechanistic systems. Complete mechanism was also restricted by the Christian belief in the soul and in free will—if nature was to be made purely mechanical, man could not be. In particular the idea that the problems of physics and other sciences were actually problems of particulate dynamics only took clear shape in the second half of the century. When this happened the science of motion had already become—or was very rapidly becoming—an exact and mathematical study, and the means were at hand for its extension far beyond the range of immediately observable bodies.

In considering the general nature of motion, what all moving things have in common and how their behaviour is to be described and accounted for, it is possible to be content with description of how things move in terms of velocity, time and distance without specially defining what is moving or what moves it. This is kinematics. On the other hand it is possible to take a wider view, to insist that the nature and size of the moving body and the cause of its motion must also be considered, and to find out (for instance) what happens when two given bodies strike each other. This leads to dynamics. In the early seventeenth century motion was treated in both these ways, but in separate compartments so that the two sets of ideas had no real connection. Such notions of a dynamical kind as there were, were qualitative only, while kinematical relations necessarily were quantitative. There was a further division cutting across these two—one of historical origin—whereby the motions of terrestrial things were entirely distinguished from those of the celestial bodies, peculiar to themselves. The great synthesis of seventeenth-century physics was the forging of these four distinct elements of science into a single unity of treatment, valid for all motions of all bodies. The first steps towards this synthesis were described in the two books of dialogues published by Galileo in 1632 and 1638.

Galileo's intellectual development was complete many years before either of them was written, and now that it is possible to trace the formation of all his ideas in some detail it is strongly possible that the content of his two great books was already largely present in his mind when he first announced his intention of writing them, in 1609. There are supporting references in the intervening years. Only his long preoccupation with astronomical observation, with the perfecting of his telescope and with the method he discovered for determining longitude at sea, and above all the effort he spent on controversy arising from his discoveries in the heavens, postponed his preparation of the substance of the two dialogues at a much earlier date. Thus the period of the formation of Galileo's scientific ideas, which is of supreme importance for understanding the transformation of physical science he effected, was over before 1630. When he wrote—initially out of a determination to relieve the Copernican theory once for all of the scepticism, misrepresentation and attacks to which it had been subject—he wrote with a full mind and long command of his materials. Until he published the *Dialogues* of 1632, however, the full scope of his thought had been hidden from all but a score or so of trusted friends and despite the acknowledgement since 1610 of his attachment to Copernicus, the immediate effect of the *Dialogues on the Two Chief Systems of the World* was terrific—in more ways than one.

The impact of the book on other than religious susceptibilities came from what was really novel in it, the group of arguments to show that the motion of the Earth is not a physical impossibility. For the most part the discussion of astronomy proper repeated what Galileo had written earlier, though the full force of his reasoning had not been so plainly brought out before. In the *Sidereal Messenger* of 1610 he had written that through the telescope one may learn with all the certainty of sense evidence that the moon is not robed in a smooth and polished surface but is in fact rough and uneven, covered everywhere, just like the Earth's surface, with huge prominences, deep valleys and chasms.

To this he added in the *Dialogues* a most ingenious demonstration that the moon could not appear illuminated as it does by reflection of the sun's light if it were formed like a convex mirror. In 1610 also he had pointed out that the discovery of Jupiter's satellites quieted the

doubts of those who, while accepting with tranquil minds the revolutions of the planets about the sun in the Copernican system, are mightily disturbed to have the moon alone revolve about the Earth and accompany it in an annual rotation about the sun ;

the same point is made again in 1632. From his *Letters on Sunspots* (1613) Galileo took the argument that the phases of Venus, revealed by the telescope, proved the revolution of this planet (and Mercury too) about the sun ; in the *Dialogues* it was expanded into a neat Socratic passage in which Simplicio was induced to draw with his own hand a pattern of celestial orbs that could only be Tychonic or Copernican. Galileo was then able to stress the greater symmetry of Copernicus' attribution of an annual motion to the orb of the Earth, between the ninth-month orb of Venus and the two-year orb of Mars. The important question of the immutability of the heavens (against which Tycho had already set himself) had also been discussed at some length in the *Letters on Sunspots* before the *Dialogues* were written, and on both occasions Galileo stated his belief that Aristotle himself would have been converted to the modern view

if his knowledge had included our present sensory evidence, since he not only admitted manifest experience among the ways of forming conclusions about physical problems, but even gave it first place.

Indeed, in these earlier tracts Galileo had several times taught his readers to expect the fuller treatment of such matters that he now gave them, while most of the astronomical arguments that did not rest on the recent telescopic discoveries had been familiar to learned men for two or three generations. The *Dialogues* were by no means the first defence of the Copernican point of view—though they have justly become the most famous, for they are the best.[5]

The reader of the *Dialogues* could not, however, obtain a very definite picture of the way in which Galileo supposed the world to be constructed. When negotiating for a post at the court of Tuscany in May 1610 Galileo had desired that he be known as Ducal Philosopher as well as Mathematician, remarking that he had studied as many years in philosophy (science) as months in mathematics. From this time Galileo was to eschew mathematics altogether, until in the years after his trial he began to write the *Discourses on Two New Sciences*. Entranced by the new prospect of the heavens, spurred on by controversy, Galileo's writings on astronomy were those of a philosopher (in his sense) not a mathematician. For the essentially mathematical attitudes of the earlier protagonists of Copernicus, maintained still by Kepler, Galileo substituted an apologia for the simple, critical elements of the heliocentric theory: (1), if there is a centre of the universe it is occupied by the sun; (2), the Earth and the planets revolve around this centre without intersection of their paths ; which turned on qualitative or physical considerations alone. Wisely, Galileo laid no stress on the claims of the heliocentric system to reproduce the observations more accurately ; nor did he say what orbits the planets followed, nor how exactly the motions according to Copernicus compared with those postulated by Ptolemy. On the contrary—adopting the philosopher's position—Aristotle rather than Ptolemy was the target of the *Dialogues*, and it was the Aristotelian rather than the Ptolemaic universe that was contrasted with the Copernican. There is every reason to believe that Galileo was a complete master of the celestial geometry of both the old and the new school ; but, for all his Platonism, he realised that descriptive analysis of the movements of the heavenly bodies could not determine their arrangement in space.

If this question was to be settled at all, it must be otherwise than by referring to *apparent* motions, or even to physical experiments on the Earth itself. Galileo knew that the celestial phenomena could be "saved"—and hence tables predicting future phenomena constructed—in countless different ways, and that no decision between them on the score of geometry was possible; he himself by turning his telescope to Venus had settled the problem of the position of her orbit which geometry alone could never have resolved. If one was aiming, as Galileo was, not merely at "saving the phenomena" by hypotheses, but hitting upon the real truth of things, then what mattered was not the Copernican system and its epicycles but the heliocentric postulate ; not mathematics but philosophy. The business of orbits was in all respects subservient to finding the pivot on which the orbits turned.

Mathematics would have been out of place in the *Dialogues*, and since this book was the result of a chain of events that began with the publication of the *Sidereal Messenger*, it was necessarily vague on the technical aspects of the heliocentric system. Galileo's attitude to Kepler is similarly explained. The former's blindness to the only astronomical discoveries of the time comparable in importance to his own—indeed, outshining his own in sheer intellectual brilliance and fortitude—has long been a puzzle. The epistolary acquaintance of the two men began in 1597, when Galileo read Kepler's *Cosmographical Mystery* ; Galileo cannot have been totally unaware of the elliptical theory of planetary motion announced in Kepler's *New Astronomy* (1609), while on the other side Kepler gave important support to Galileo's observational discoveries. Pleased by this timely support, when all the world was incredulous, Galileo nevertheless regarded Kepler as a dangerous ally. Two years after Kepler's death (1630) he wrote :

I doubt not but that the thoughts of Landsberg and some of Kepler's tend rather to the diminution of the doctrine of Copernicus than to its establishment, as it seems to me that these (in the common phrase) have wished too much of it . . .

And he told another friend " I always value Kepler for his subtle free-ranging genius (perhaps too free) but my philosophy is very different from his ".[6] It would have been more exact perhaps to say that he did not share Kepler's metaphysics ; at any rate, there is no need to postulate anything but intellectual unlikeness to explain the lack of cordiality between the two astronomers. If Galileo can be forgiven for missing the (then) relatively unimpressive point of Kepler's planetary theory that the sun is at the focus of each elliptical planetary orbit—so making the sun the genuine pivot of the whole system in a way that it was not in Copernicus'—then he was perfectly correct in supposing that to replace circles by ellipses proved nothing. From Galileo's point of view Kepler's discoveries had not at all eased the burning question of early seventeenth-century science. They were rather another version of the kinematical illusion from which philosophy must turn its attention. He is not to be blamed for failing to find—what he could not have found—the dynamical justification of the heliocentric system which Newton discovered in Kepler's laws.

As a result, the *Dialogues* portray a heliocentric theory without a planetary system. The book assumed that the Earth and planets move concentrically in perfect circles about the sun, as in the common simplified picture of the Copernican universe. It made no mention of the epicycles and eccentricities that Copernicus had had to build into it in order to make it fit observation. To this extent the *Dialogues*, ignoring all the difficulties of planetary kinematics to which Kepler had already presented a solution, were a poor guide to real astronomy, and expound a mere oversimplified travesty of the truth to whose defence they were devoted. But, if it is remembered that what Galileo was in fact defending was the heliocentric concept, and that his attitude was philosophical not mathematical, this criticism carries no weight. Galileo was well aware that perfect circular motion would not work astronomically ; what mattered to him was its philosophic import. Epicycles and ellipses made no sense to him philosophically, whereas motion in circles did :

it seems to me [he wrote] one may reasonably conclude that for the maintenance of perfect order among the parts of the universe, it is necessary to say that movable bodies are movable only circularly ; if there are any that do not move circularly, these are necessarily immovable, nothing but rest and circular motion being suitable to the preservation of order.[7]

So spoke Salviati, and there is little cause to suppose he was not (as in other passages where his words are more true) speaking for Galileo himself. The idea is false, and it is easy to remark that Galileo was prostrating himself before the ancient assumption of the perfection of uniform circular motion. Kepler had freed himself from this dogma, Galileo not. Observe the other side, however ; Kepler's ellipses drew him into the philosophical horrors of magnetic mysticism, and these Galileo escaped. One cannot take more than a single step at a time, and no one could yet see that Kepler's step forward in the mathematics of astronomy was, after all, absolutely compatible with Galileo's different step forward in its philosophy.

As might be expected, the physical nature of the universe of the *Dialogues* is much clearer and more advanced than its kinematics. Galileo did not reject the reality of the celestial orbs as vocally as Tycho and Kepler had done, but he was of their mind. In a letter of 1612 he wrote :

But that Nature has need of solid eccentric orbs and epicycles to effect such movements [as those of the planets] is I think a mere piece of fantasy, and a needless chimera.[8]

Indeed, he (mistakenly) attributed the same opinion to Copernicus himself. For him, the planets like the Earth and moon shone by reflecting the sun's light. All these bodies were Earth-like spheres, made of the same kind of material, rising into mountains and falling into valleys, actually visible on the moon. At first Galileo believed that as he detected seas on the moon, so it was also surrounded by an atmosphere ; but by 1632 he had rejected the atmosphere and was no longer certain of the seas. And very logically he pointed out that no living form imaginable by man could inhabit the lunar surface.

The celestial bodies moved freely in space—Galileo did not declare what, if anything, filled the celestial regions—because once moved there was nothing to resist their motion. Impelled at the creation Earth and planets revolve for ever. According to an inaccurate calculation he made, if all the planets had been created at one place and thence fallen freely towards the centre of the universe, each would at its due distance from the centre have acquired the velocity with which it describes its orbit.[9] Beyond Saturn lay the stars, " so many suns " ; like every Copernican Galileo knew that in comparison with the remoteness of the stars the size of the solar system was minute. In the *Dialogues* he approved the limitation of the stars within two spherical boundaries, beyond which it was impossible that space should extend, though he also denied that it was possible to *prove* whether the universe was infinite or finite, of a definite shape or not. The idea that the stars are embedded in a solid sphere was of course an absurdity to him ; and the revelation by the telescope of innumerable stars invisible to the unaided eye suggested not that such stars were too small to be seen normally, but that they were too remote. Hence the stars must vary greatly in distance and Galileo's true opinion seems to have been that the furthest stars might be indefinitely (though not infinitely) remote. This was a question that neither observation nor logic could settle.

Most of the remaining astronomical discussion in the *Dialogues* contains no original views or novel arguments in favour of the Copernican universe. The effect of the book lay in its bringing to bear, with great force, the lessons of the recent astronomical discoveries and the arguments long developed by the Copernicans against the common prejudice in favour of the Aristotelian cosmography, in a manner accessible to all readers. Many of Galileo's earlier writings had been printed in small editions, scarcely obtainable even in Italy, and much of the Copernican defence had been offered in mathematical treatises that only specialists could comprehend. Creating the character of Simplicio, Galileo addressed the scholar ill at ease in geometry, while the *Dialogues* in the Latin translation of

Matthias Bernegger (1635) spread rapidly all over Europe, despite the Church's ban on the Italian original. They were even translated into English in the same year, as Thomas Hobbes informed Galileo, but no English version was actually available until 1665. At the very moment when many mathematicians and other devotees of science were relenting towards the Copernican theory—and this was happening among the Italians, and even among the Jesuits—Galileo's book appeared with the most eloquent and persuasive justification of it that was ever written. And after the famous trial *not* to be a Copernican was to countenance popish obscurantism.

The central theme of the *Dialogues* was not astronomical at all, however ; it was physical. The discoveries that Galileo announced in them were those of mechanics, not of new spectacles in the sky. The reason is well known. Aristotle's physics and his cosmology alike started from the assumption of the fixed, central Earth and there could be no destruction of his cosmology without a simultaneous destruction of his physics. Now, to replace Aristotle's cosmology there was at hand the universe of Copernicus, rendered more substantial by the recent telescopic revelations ; a new view was certainly not lacking. But before Galileo there was nothing to take the place of Aristotle's physics. Every critic of Copernicus had rejoined that while it was imaginable that the sun and stars should be motionless, it was impossible, manifestly counter to everyday experience, that the Earth should move. Everyone would echo Simplicio's words, "The crucial thing is being able to move the Earth without a thousand inconveniencies"—inconveniencies arising from the entire disruption of the familiar, stable world of experience. After all the question of whether the sun moves or not turned out to depend on what happens in carriages and boats, or to balls rolling down inclined planes. This was made clear soon after the opening of the *Dialogues* when Galileo drew attention to the Aristotelian doctrine of the centre of the universe, to which not merely the celestial motions but all motions were to be referred. If this doctrine were maintained, then indeed the Earth could not move ; Galileo's task was to prove that everything would happen as it does without such a fixed point of reference, so that the Earth could move. Or, more generally, he had to show that ordinary motions, like those of arrows or balls, were *more* intelligible than before if the doctrine of the fixed centre were dispensed with, not less so.

In accomplishing this colossal task, the foundation of all subsequent physical theory, Galileo admittedly derived aid from his predecessors in dynamics and kinematics. From the sixteenth-century notion of impetus he went forward to his own completely novel conception of inertia. From the " Merton Rule " * he went forward to his own theory of free fall. The *Dialogues* represent Galileo's thought at the end of this process of development and offer his main statement of the unity of the science of motion in the sky and on the Earth, as against the dichotomy of Aristotle.

The matter is clearly put by Salviati :

Now as to Simplicio's final remark. He says that it is folly to debate whether parts of the sun, moon, or any other celestial body, if separated from their whole, would naturally return to it; because (he says) the case is impossible, it being clear from Aristotle's proofs that celestial bodies are unchanging, impenetrable, indivisible etc. I answer that none of the conditions by which Aristotle distinguishes celestial from elemental bodies has any other foundation than what he deduces from the difference in natural motion between the former and the latter. In that case, if it is denied that circular motion is peculiar to celestial bodies, then one must choose one of two necessary consequences. Either the attributes of generable-ingenerable, alterable-inalterable, divisible-indivisible, etc. suit equally and commonly all bodies in the world, as much the celestial as the elemental, or Aristotle has erroneously deduced, from circular motion, those attributes which he has assigned to celestial bodies.[10]

* See p. 67.

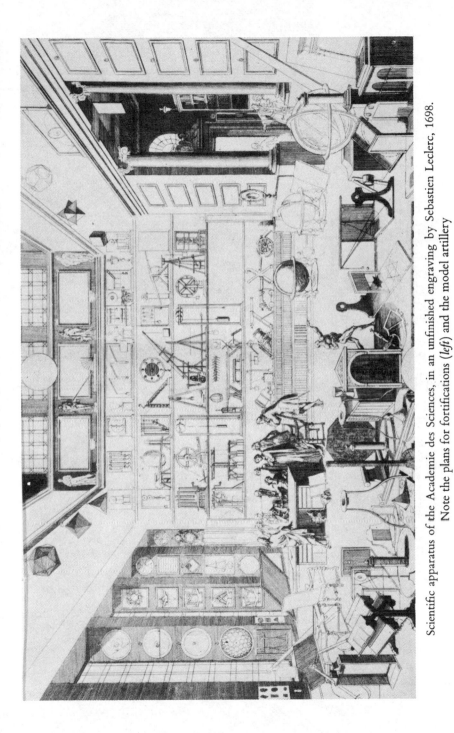

Scientific apparatus of the Academie des Sciences, in an unfinished engraving by Sebastien Leclerc, 1698. Note the plans for fortifications (*left*) and the model artillery

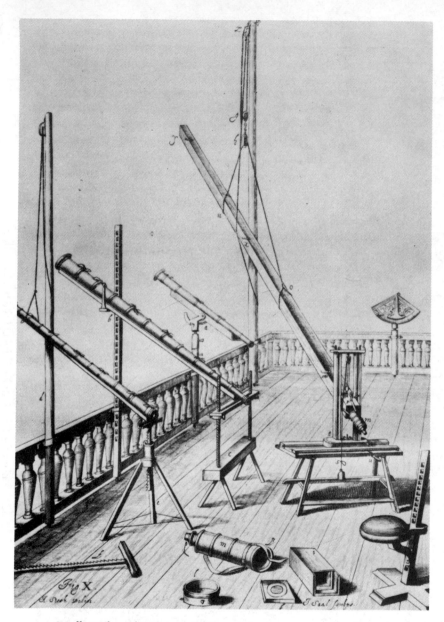

Small- and medium-sized astronomical telescopes, c.1660. *k* is an
objective and *m* an eyepiece. From Hevelius's *Machina Coelestis*

In other words, once it is admitted that the science of motion is common to heaven and Earth, then it follows that all antitheses drawn in science between the celestial and terrestrial worlds are false ; there is only one body of knowledge that is applicable to the whole realm of nature.

The complete unity of the science of motion was not to be gained at one stroke, however. For just as Aristotelian physics was based on the classification of movement into celestial and terrestrial, natural and violent, so in the *Dialogues* Galileo clung to a distinction between naturally-occurring motions which are uniform and circular, and forced ones which are accelerated and rectilinear. It was still necessary to find something unique in the eternal, changeless rotation of the skies.

It is twice asserted by Salviati that

if all integral bodies in the world are by nature movable, it is impossible that their motions should be straight, or anything else but circular ; and the reason is very plain and obvious. For whatever moves straight changes place and, continuing to move, goes even further from its starting-point and from every place through which it successively passes. If that were the motion that properly suited it, then at the beginning it was not in its proper place. So then the parts of the world were not disposed in perfect order. But we are assuming them to be perfectly in order ; and in that case, it is impossible that it should be their nature to change place, 'and consequently to move in a straight line.

Hence the most that can be said for rectilinear motion is

that it is assigned by nature to its bodies (and their parts) whenever these are to be found outside their proper places, arranged badly, and are therefore in need of being restored to their natural state by the shortest path.[11]

Accordingly it is perfectly in harmony with the structure of the universe that the Earth should move in a circle, like the other similar planetary bodies, and if there is a centre of the universe (or rather, the solar system) it is to be found in the sun, not the Earth. Bodies falling back to the surface of the Earth are by no means seeking that centre ; instead they are moving so as to unite with the whole Earth, by their natural tendency to form and preserve it. Even these falling bodies do not really, but only to appearance, move in straight lines and thus they confirm the universal principle.

Here, at the very beginning of the *Dialogues*, is encountered Galileo's conception of uniform, inertial motion in a circle. It is a false conception, but the only one permitting him at this moment to fit both terrestrial and celestial motion into a single frame of ideas. If his concept of inertia had been more correct he would have needed to discover the whole Newtonian system—or done nothing at all. In order to make the planetary system gyrate inertially like a free-spinning wheel he had to use circular inertia to get rid of the Aristotelian spokes and felloes. For once, instead of cosmology being explained in terms of mechanics, cosmology is forcing its requirements upon mechanics. But—as Galileo showed later—he did not forget circular inertia when he came down to Earth, for he took care to point out that the difference between straight lines and vast circular arcs was imperceptible.

In the second day of the *Dialogues* Galileo turns to the question of what happens when a stone falls from the top of the mast of a ship in motion or at rest, an analogy illustrating the problem of fall on a moving or motionless Earth. Will the stone fall at the foot of the mast in either case, or, as the Aristotelians claimed, would it be left far behind the moving ship in its descent ? In answer, Galileo argued first that a ball on a horizontal plane, free from all impediments, would continue to roll indefinitely without losing any velocity imparted to it ; and that the surface of the Earth forms such a continuous plane. Next, he asserted that the stone moving with the top of the mast will continue to move with the same velocity over the surface beneath when liberated, as though rolling on a plane, replying to Simplicio's accusation that he was thus conferring upon the stone an " impressed " force (or impetus) with just the arguments that the fourteenth-century philosophers had used in support of the idea of impetus. Finally, following Tartaglia's idea of a " mixed motion ", Galileo denied that the fall of the stone towards the Earth would in any way interfere with its progressive motion. Hence, thanks to its inertia, the stone fell at the foot of the mast when the ship was under way no less than when it was motionless, and the same law applied to the moving Earth. But Galileo's inertia preserved circular motion. As the great circle round the Earth was to Galileo an endless plane—true enough for small-scale experiments—so the path which the stone faithfully follows is a circular path. For in Galileo's reasoning everything happened as though the ship were at rest ; that is to say his (perfectly correct) intention was to maintain that in an inertial system there is no relative displacement of bodies. What he overlooked was the fact that the ship and the stone do not form an inertial system, because they are *constrained* by gravity to follow the Earth's curvature. The very need to demonstrate the (approximate) truth that in the ordinary movements of bodies at the Earth's surface their inertia completely masked any effect due to the rotation of the Earth, induced Galileo to believe that inertia caused them to move in paths parallel to its spherical surface.

The experimental error is negligible of course, but the conceptual error is not. Inevitably it was again one that rendered Galileo's arguments more straightforward than they would otherwise have been, had his understanding of inertia been fully correct. His position throughout the physical discussions of the *Dialogues* was firmly put by Salviati :

> I might add that neither Aristotle nor you (Simplicio) can ever prove that the Earth is *de facto* the centre of the universe ; if any centre may be assigned to the universe, we shall find the sun to be placed there . . .[12]

This opinion rested wholly upon Galileo's concept of inertia. When inertia is correctly defined there is a possibility—as Newton was the first to point out in 1679—of determining from the descent of bodies whether the Earth moves or not (though the experiment is not a good one to choose). It is a curious irony that an imperfect concept of inertia enabled Galileo to refute the dynamical absurdities of his geocentric opponents, while an accurate one would (in theory) have permitted that test of Copernicanism that he denied was possible in principle. One step at a time. . . .

Galileo, it has been truly said, " geometrised space ". The phrase epitomises many aspects of the *Dialogues* : Galileo's renunciation of a privileged centre for the universe, and of all that hung upon it in Aristotelian physics ; his abstraction from the incumbrances of experimental reality, such as the limits of planes, the friction of rolling balls, and so on ; and his kinematical treatment of moving bodies, taken further in the *Discourses on Two New Sciences* (1638). In fact the generalisation of the science of motion effected by Galileo necessarily involved the geometrisation of space. Yet at the same time it is in the *Dialogues* a geometry of circles rather than straight lines ; in the *Discourses* indeed it is only of straight lines because Galileo there confined himself to the small-scale. In the *Dialogues* he could not do so because he was concerned with the spatial geometry of the universe, so, correspondingly, the kinematics in that book is one of curvilinear motions. If there is a fault in Galileo's science of motion, then—the fault to whose detection and correction so much of the physical theory of the later seventeenth century was devoted—it is one that arises from the geometrisation of space done in terms both of circles and of straight lines. Galileo, who had not clearly seen the incompatibility, left its resolution to his successors.

To us, the incompatibility is evident in the minor problem that Galileo raised of what would happen if a stone could fall right down to the centre of the Earth. Its apparent path is obviously a straight line. But how would it be seen to move by an observer not turning with the Earth ? Galileo was well aware that on a flat surface a projectile describes a parabola, and he had also reasoned that a stone carried round with the speed of the Earth's rotation, was itself a kind of projectile. Yet in the *Dialogues* he failed to couple these ideas. Although clearly at least the start of the fall of this peculiar stone is a normal terrestrial event he looked at this motion towards the centre in cosmic terms and endowed the stone with a circular inertia. It would, he suggested, describe at a uniform speed a semicircle on the Earth's radius, so giving an observer standing on the Earth the illusion of straight-line, accelerated fall. Later Galileo qualified this notion as a mere piece of fancy, but fanciful thoughts are revealing. Here in an artificial problem touching both terrestrial and celestial mechanics Galileo slipped in notions of circular inertia and of the desirability of making all natural motions circular that at once conflicted with the plane geometry notions of the later *Discourses*.

There is one case so counter to a theory of inertial motion in circles that it was seemingly impossible for Galileo to disregard it. The potter's wheel, it was well known, flung off bits of clay and even a heavy stone would fly from a wheel spun fast enough. This was indeed, precisely the traditionalist objection to the rotation of the Earth : everything would be thrown up into the air. Salviati brought forward this very difficulty :

If, then, the Earth were to be moved with so much greater velocity [than wheels], what weight, what tenacity of lime and mortar, would hold rocks, buildings and whole cities so that they would not be hurled into the sky by such precipitous whirling ?[13]

Naturally he also presented Galileo's answer to it. In spite of its geometrical ingenuity it is not a very good answer, for it proves too much. According to Galileo's reasoning centrifugal force can never remove something from a circular motion, if there is a contrary force towards the centre, however weak. For, examining the angle of contact, it appeared to Galileo that as the tangent along which the body would be projected approximates more and more closely to the circle in which it moves, the ratio between the length of the tangent and that of the secant grows greater and greater (figure 1).

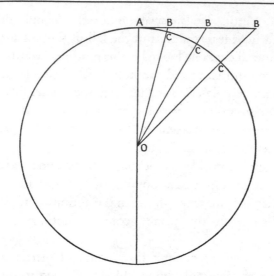

FIG. I.   As the angle AOB diminishes, BC diminishes in proportion
to AB

However great the force acting along the tangent AB, it can never move the body since only a minute retaining force is required to pull it back along the infinitely less distance of the secant BC. This argument is false because the centrifugal force in a given circular motion increases as the square of the velocity ; the Earth could indeed project rocks into space, if it moved fast enough. On Galileo's reasoning its inability to do so is independent of its speed. Finally, Galileo permitted Sagredo to observe :

> thus it might be supposed that the whirling of the Earth would no more suffice to throw off stones than would any other wheel, as small as you please, which rotates so slowly as to make but one revolution every twenty-four hours.[14]

Another mistake, since if angular velocity is constant the centrifugal force increases with the radius. Its source in Galileo's thought was identical with the source of his error with regard to rotary inertia. The larger the radius of the circle, the closer its circumference approximates to a straight line ; if the moving body has no tendency to deviate from the straight line, its tendency to deviate from the circle should grow less as the arc becomes more nearly straight.

Starting from the logical (and verifiable) deduction that a stone flies in a straight line from a string, Galileo had ended with a theory of rotational force consistent with his belief in the rotation of the Earth and in rotary inertia too : for to say that the buildings on the Earth cannot be flung into space is to give them rotary inertia. That he had ended with a theory that would also prevent the stone leaving the sling if the least force held it back seems to have escaped him. How could this happen ? The answer seems to be in Galileo's manner of reasoning. This did not explore the dynamics of the situation. The only way to handle the problem of centrifugal force in connection with the rotation of the Earth, and to settle it, was to calculate the force arising from the Earth's spin and compare it with the force of gravity. It is no detriment to the fame of Galileo that he could not do this, that he did not anticipate the discoveries of Huygens and Newton half a century later. That he could be satisfied by a specious solution arose from his taking a kinematical approach to be quite adequate ; he compared not forces but the lengths of lines. For Galileo the stone in the sling was a convenient device rather than a means of experiment, and once it had evoked the proper response from Simplicio it was set aside and there was no further reference to experiments. *A priori* geometrical reasoning seemed to make experiment superfluous.

Once before Simplicio had been thrust into an empiricist role and Galileo had gone on more openly to refute its validity. When the question of the fall of the stone on board ship was debated Salviati had hinted that his conviction that it would fall to the very foot of the mast was based not on trial but on reason ; to which Simplicio rejoined :

So you have not made a hundred tests, nor even one ? And yet you so freely declare it to be certain ? I shall retain my incredulity, and my own confidence that the experiment has been made by the most important authors who make use of it, and that it shows what they say it does.[15]

Nullius in verba; Bacon and the Royal Society would have put it better, but they would have sympathised with Simplicio's indignation. They would have been wrong and Galileo right, for thought not experiment formulates the concepts of a new science. But Simplicio was not allowed to refer again to the experimental test that alone reveals whether, in the world of things, that application of concepts is just or not. Galileo felt he did not need to make the crew of a galley strain at their oars to prove his concept of inertia; because he had penetrated the nature of motion he knew what must happen. It could not be otherwise, and the later experiments of others proved it was so. If he failed similarly to solve the problem of centrifugal force it was because the application of his study of motion was overextended and even he could not make kinematics do the work of dynamics.

If we except astronomy, Galileo's fame was not to rest chiefly upon his activity as an experimentalist. It is true that he was experimentally minded, that he was ready to entertain new ideas and to explore the possibilities of new means of acquiring knowledge of nature, but that is a different thing. It is also true that he had an eye open for original observations that provoked theoretical questions, as did for him the information that water does not rise higher than 32 feet in suction-pumps, but again it is the theory that he pursues rather than further experimental inquiry. Only a few genuine experiments that Galileo made are recorded and in all his letters there is very little mention of actual experiments. He did indeed spend much time on the perfection of instruments such as his proportional compass, his thermoscope, and above all the telescope. This was technical development. On the contrary, apart from his astronomical observations Galileo was great as a theoretical scientist. To arrive at the concept of inertia, even with the imperfections of Galileo's formulation, was a colossal achievement. As for experiment to verify this concept, Galileo brushed it aside as irrelevant. One needed a feat of imagination to accept this new idea, which was never perfectly demonstrable from the world of experience, and therefore he insisted that if we apply our imaginative insight correctly to our ordinary experience of the world, we cannot fail to perceive that the law of inertia holds. In a sense experimental details would only confuse the intellect which alone is capable of apprehending the truths of nature beneath the deceptive surface of things. Of this there could be no better example than the heliocentric theory itself, founded on the belief that the "experimental" stability of the Earth is an illusion, and appearing to deny that

> the senses and experience should be our guide in philosophising. [For] in the Copernican position, the senses much deceive us when they visually show us, at close range and in a perfectly clear medium, the straight perpendicular descent of very heavy bodies. Despite all, according to Copernicus, vision deceives us in even so plain a matter and the motion is not straight at all, but mixed straight and circular.[16]

No one could be both a Copernican and a simple-minded empiricist, when the motion of the Earth was only to be discovered by employing reason to dissolve its seeming fixity. Hence Galileo employed the Socratic method, constantly assuring Simplicio that he knew the truth already, if only he could be brought to view it squarely, without prejudice or preconception, and with due attention to the principles of geometry. When erroneous beliefs—such as the Aristotelian ideas of motion—arose not from want of evidence but from a habitual failure to see the evidence in the right light, it was a change of vision that was required, not still more evidence. Here Galileo's position is really much closer to Aristotle's than it might seem at first sight. For Galileo nowhere in the *Dialogues* maintained that the strength of his concept of inertia rested upon its agreement with experiments ; at the most, experience could provide ocular evidence of its validity, at the worst it could be shown that no experience was against it. Thus, whatever their illustrative value, there is no need to discuss as experiments what Galileo had to say about the motions of arrows, bowls thrown in sport, or balls dropped from carriages. A passage more central to Galileo's theory of inertia is that in which he analysed the descent of bodies in, for example, the cabin of a ship. Once more the cabin forms a (circularly) moving frame of reference, and again everything happens as though this frame were at rest. Therefore it is obvious that

> that motion which is common to many moving things is idle and inconsequential to the relation of these movables among themselves, nothing being changed among them ; and that it is operative only in the relation that they have with other bodies lacking that motion, among which their location is changed.[17]

This is a manner of expressing the law of inertia most serviceable to the argument of the *Dialogues*, where Galileo's eye was on the properties of inertial systems—like the Earth—not on those of single isolated bodies. Again there is no need to go to sea to make the experiment, since no one would dispute the facts.

The classic story of Galileo's defeat of Aristotle by dropping balls of iron and lead from the Leaning Tower of Pisa in the presence of a multitude of astounded scholars is now credited by no one. Such an experiment would not have proved what it was supposed to prove nor had Galileo any reason to make it once he possessed the reasoning that taught him that Aristotle's law of fall was false. Really the ancient problem of the speed of falling bodies was irrelevant to the theme of the *Dialogues*, but Galileo found a way to bring it in through an absurd statement of the anti-Copernican Locher made nearly twenty years before. Accordingly it was in 1632 that Galileo first published his contention that in uniformly accelerated motion (1) the instantaneous velocity (v) is proportional to the time elapsed (t) ; (2) the distance traversed (s) is proportional to t^2 ; and (3) $vt = 2s$. The first two propositions were given without proof, the third—already perceived in the Middle Ages— was demonstrated geometrically from the second. Of course

Galileo was also careful to assert (without proof) that the free fall of heavy bodies is a case of uniformly accelerated motion.[18] Here again the law proposed is a universal law ; it applies not merely to free fall but to every example of uniform acceleration ; and it excludes the possibility that the rate of fall varies with the density of the body.

No doubt Galileo's discoveries in kinematics were less utterly original than he thought. Yet as far as can be discovered this bare announcement of 1632 occasioned wide admiration, if not always approval, and no one looked back to the anticipations of Swineshead or Oresme or even of Dominico da Soto less than a century before. Europe waited impatiently for the full explanation that came in the *Discourses on Two New Sciences* six years later. Galileo's friends broke into expressions of hyperbole. One of them, Fulgenzio Micanzio, wrote to him soon after reading the *Dialogues* :

An hour seems to me a thousand years in waiting to see the other dialogues, as I am sure we shall find in them part of what was promised about natural motions and projectiles . . . I delight myself with these and long to see this work of yours done, which surely no one hopes for more than I.

Micanzio's tedious hours stretched on for two and a half years until, when he did see part of the manuscript of the *Discourses*, he wrote (January 1635) that he had

looked through them with supreme avidity and enjoyment. It's amazing how you observe the effects of nature in trivial everyday things before everyone's eyes, and rise to profound speculations, deduced from true principles that satisfy the mind.[19]

The design for a treatise on " two new sciences "—the one dealing with the strength of materials, the other with the motion of projectiles—had formed in Galileo's mind as early as 1609, when he described it in a letter to Antonio de' Medici.[20] References to his new theory of motion were fairly common in letters between Galileo and his friends during subsequent years ; Cavalieri, for

instance, received with delight in 1630 the news that Galileo had
turned his thoughts to motion again, " seeing that with this science
and mathematics joined together we can prepare ourselves for
speculations about natural things, and with great confidence hope
for the desired knowledge ".[21]   At this time, though his capacity
for work was already reduced by age and ill-health, Galileo hoped
to continue the promised discussion of motion from the *Dialogues on
the Two Chief Systems of the World* in another book to follow very
soon after.   In fact, not only was the completion and publication of
the *Dialogues* delayed, but its consequences kept Galileo from
writing for longer still.   He was dismayed at his fate, and feared
that the censorship would never allow him to print again.   By the
end of 1634, however, Galileo had begun to put together " what
remains of my labours, which is what I value most for it is all new
and all mine,"[22] and the new book was finished in the next year.
After a licence to print in Venice had been refused there were
further long delays before the volume at last appeared in July 1638,
and still another year passed before a ship at last brought a copy from
Leiden to Galileo's hands.   He could no longer see it.   Yet even
when sight had gone

> and consequently the power to investigate more profound
> propositions and demonstrations than the last discovered and
> written by me, I have passed the hours of darkness in the night
> in going over the first and most simple propositions, re-
> arranging and disposing them for better order and force . . .[23]

With indomitable spirit Galileo still proposed to embellish his last
great work,

> if I have enough strength to improve and enlarge my former
> writings and publications on motion, adding some speculations
> to them and especially those respecting the force of percussion,
> in the investigation of which I have spent many hundreds and
> thousands of hours and finally reduced it to a fairly simple
> explanation, so that others can understand it less than half
> an hour.[23]

The expectations of Galileo's friends in Italy and abroad were not defrauded. Cavalieri expressed the sentiments of them all :

I believe it can be said with full reason that, taking sound geometry as your guide and with the fair wind of your high genius, you have successfully negotiated the immense ocean of indivisibles, vacuum, infinities, light and a thousand other difficult, nay extraordinary matters, each one of them capable of shipwrecking a man of great wit. . . .[24]

He was right. The *Discourses on Two New Sciences* did touch on almost every aspect of physics, yet Cavalieri did not mention the outstanding feature of the book, its treatment of motion, perhaps because he had unwittingly offended Galileo by rushing into print on this topic himself some years before. The solid theoretical part is found largely in the Third and Fourth Days of the dialogue and seems to have been transcribed with little change from some earlier draft of Galileo's, for the conversation that was almost as lively during the first two Days as in the *Dialogues* of 1632 is virtually abandoned, leaving Sagredo and Simplicio mute listeners. In the First Day, however, before resorting to kinematics Galileo developed at length his argument (already broached in the earlier book) that physical bodies do not fall to the ground at speeds proportional to their densities. Far otherwise :

the variation of speed observed in bodies of different specific gravities is not caused by the difference in specific gravity but depends on external circumstances and, in particular, upon the resistance of the medium, so that if this is removed all bodies would fall with the same velocity.*

If, Galileo asserted, experiment showed that a 100 lb. weight struck the ground a little ahead of a 1 lb. weight, this did not excuse the gross error of Aristotle in thinking that the former would fall

---

* One may reflect that all the same the less density is the cause of slower fall. Galileo meant, of course, that change in density is not a self-sufficient cause of variation in speed of fall ; the resistence of the medium is equally a cause of this effect.

100 feet while the latter fell one.[25] And he was careful, in analysing the effect of the resistance of the medium in retarding motion, to discriminate between variations arising out of differences in density, and those arising out of differences in surface/weight ratio.

The argument for the nearly uniform acceleration of dense falling bodies was beautifully put in one of those neat dialectics so typical of Galileo's scientific argument. Suppose Aristotle to be right : tie a light stone (falling speed 4) to a heavy stone (falling speed 8). What will be the speed of the two together ? Let the larger stone be tied above the smaller one : then the larger will be so slowed down by pushing the smaller that the speed of the whole mass will now be less than 8. Yet the double stones are heavier than either singly, and should fall faster than 8. Let the small stone be tied above the larger; just the same happens because the small stone cannot weigh down on the larger which (*ex hypothesi*) tends to travel faster. Which is absurd. The only solution is to agree that large and small (of the same density) fall at the same speed. From such passages it could be argued that Galileo's logical weapons were no less forceful than any experimental ones. But, if the proposition that all bodies fall at the same speed is only strictly true of the vacuum, a fresh issue of principle seems to be raised. Aristotle had not merely argued that a vacuum in nature is impossible but that (if one were forced by some means) bodies would not move in it as they normally do. Could any statements made about motion in the impossible condition of vacuous space be relevant to the actual motions of real bodies in air ? Once more there is a covert challenge from naive empiricism. Galileo certainly believed that there are minute vacuities between the solid atoms of matter, yet no more than Aristotle did he suppose that there are vacuous spaces in nature of considerable size " though they might possibly be produced by violent methods, as may be gathered from various experiments ".[26] Observation shows, however, that differences in falling-speeds diminish as the density of the medium grows less ; by inference if the resistance were zero the difference would be nil. The significance of the unattainable vacuum for Galileo was that he could state for it, in complete confidence, the unverifiable law of uniform acceleration. And thus, by allowing for the effect of the distinguishing condition, air resistance, the varied speeds of actual bodies really falling could be accounted for. In other words : only by imagining an impossible situation can a clear and simple law of fall be formulated, and only by possessing that law is it possible to comprehend the complex things that actually happen. Idealism (or abstraction) is *not* delusion because it ignores the complexities and discrepancies of reality ; on the contrary, only through idealism can the reality explaining the complexities and discrepancies be discerned. Then, " facts which at first sight seem improbable will, even on scant explanation, drop the cloak which has hidden them and stand forth in naked and simple beauty ".[27]

The supreme instance of idealism or abstraction in scientific method is the use of mathematics, especially (in Galileo's time) geometry, for the study of physical events. If the non-entity of the vacuum is inaccessible to experience so are perfect spheres and perfect planes. The fact presented Galileo with opportunity for his most outright justification of idealism. Challenged by Simplicio in the *Dialogues* :

> After all, these mathematical subtleties do very well in the abstract, but they do not work out when applied to sensible and physical matters ;

Salviati had responded

> when you want to show me that a material sphere does not touch a material plane in one point, you make use of a sphere that is not a sphere, and of a plane that is not a plane.

But, he went on, this is a case where geometry equally provides the same answer : even in the abstract an imperfect sphere does not make point contact with an imperfect plane.

> So that what happens in the concrete up to this point happens in the same way in the abstract. It would be novel indeed if computations and ratios made in abstract numbers should not

thereafter correspond to concrete gold and silver coins and merchandise. Do you know what does happen, Simplicio? Just as the computer who wants his calculations to deal with sugar, silk and wool must discount the boxes, bales and other packings, so the mathematical scientist, when he wants to recognise in the concrete the effects which he has proved in the abstract, must deduct the material hindrances, and if he is able to do so, I assure you that things are in no less agreement than arithmetical computations. The errors, then, lie not in the abstractness or concreteness, not in the geometry or physics, but in a calculator who does not know how to make a true accounting.[28]

Thus a physical law holding absolutely in unreal conditions is like a geometrical theorem which is only true for geometrical bodies. The law of fall is not falsified because air is not a vacuum, any more than geometry is falsified because a ruler is not absolutely straight.

Now at last, in the *Discourses*, Galileo proposes a definite experiment to vindicate his doctrine. Take two equal pendulums, one with a bob of cork, the other with a bob of lead. Set swinging, their arcs whether of unequal or equal amplitude are always performed in the same time. Perhaps this should be called a demonstration rather than an experiment; in any event (since the cork will be brought to rest before the lead) it does not prove the law of inertia. The real interest of this experiment is that it introduced the pendulum into science. Galileo saw that the swinging of a pendulum was an example of harmonic motion and showed how the musical harmonies between vibrating strings could be represented by the oscillations of pendulums whose cords were as 16, 9 and 4 units. He asserted—and satisfied himself by trials—that the time of swing of any pendulum is independent of its amplitude, being proportional to the square root of its length. The reasoning underlying these statements can be deduced from what is said in the Third Day of the *Discourses*. Though neither was completely demonstrated, the achievement is remarkable. First Galileo proved that the time of descent along any chord drawn from the base of an upright circle is the same as that of a fall along the perpendicular diameter (figure 2). Next, that the descent to the base along any arc of less than 90° is faster than that along the corresponding chord (Galileo mistakenly thought that this implied that such a circular arc provided the quickest route of descent between any two points).

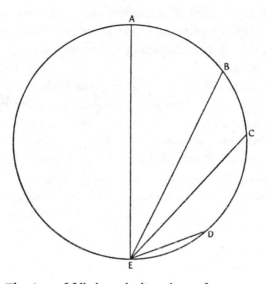

FIG. 2. The time of fall along the lines drawn from A, B, C, D, etc. to E is the same

From which reasonably—but erroneously—he inferred that all descents along any arc take the same time, and hence that the swings of the pendulum must be isochronous. Consequently it followed that as the time taken to descend the diameter is as the square root of its length, the time for descent along any arc, or the swing of the pendulum, must be in the same ratio.

The theory of the pendulum, consisting of the refinement and extension of Galileo's results, was a nodal point in seventeenth-century dynamics. For his successors, as for Galileo himself, its requisite assumption was the conservation of momentum, a truth evident in the ascent of the pendulum bob to the level from which it fell (approximately). As Galileo expressed it :

in general every momentum acquired by fall through an arc is equal to that which can lift the same body through the same arc.[29]

This was a wholly novel principle of mechanics. Galileo did not regard it as rigidly established, but as made highly probable by the pendulum experiment. He preferred, however, to state the conservation of momentum in a different form applicable to rectilinear motion :

The speeds acquired by one and the same body moving down planes of different inclinations are equal when the heights of these planes are equal.

This he called an " assumption ". If it were false, then either momentum is not conserved (so that a ball would not roll up a plane, excluding friction and so on, to the height from which it fell) or it would be possible to raise a ball indefinitely by running it up and down suitable planes. Galileo rejected both these possibilities and no doubt (like Sagredo) looked upon the conservation of momentum as a self-evident principle. But he could not prove it directly, any more than the principle of inertia could be proved directly.[30]

Besides this assumption Galileo postulated for his kinematics a famous definition of uniform acceleration :

A motion is said to be equally or uniformly accelerated when, starting from rest, its momentum receives equal increments in equal time.[31]

Of course, for momentum Galileo could as well have written, and does in fact mean, " speed ". The definition was not new ; what was new was Galileo's conviction that it was the only possible definition, the purely kinematic status he gave to it, and the uses to which it was put.

Galileo, like his predecessors for three centuries, had thought that the speed of a falling body must be proportional either to the distance it had fallen from rest, or to the time elapsed. In 1604 he had regarded the former alternative as the more probable, by 1611 he had discovered (as he thought) a proof that the second was correct.[32] This proof appears in the *Discourses* and makes use of the " Merton Rule ".* Suppose that the velocity increases with the distance fallen ; when the body has fallen 4 feet the speed is say 10, at 8 feet the speed is 20, at 12 feet 30 and so on. Then by the Merton Rule the times taken to fall 4, 8, 12 . . . feet are as $\dfrac{4}{10 \times \frac{1}{2}}$, $\dfrac{8}{20 \times \frac{1}{2}}$, $\dfrac{12}{30 \times \frac{1}{2}}$ . . . that is always 4/5 units. Which is impossible unless the fall is instantaneous. Galileo was right in rejecting the simple proportionality between speed and distance, though his belief that it led to a *reductio ad absurdum* was open to objection. It may be regarded as a happy short cut which confirmed for him the "intimate relationship between time and motion" that was the foundation of his kinematics.

He did not attempt to derive this relationship from a causal account of the acceleration of falling bodies, describing the explanations of philosophers as fantasies over which it was futile to linger :

At present it is the purpose of our author [himself] merely to investigate and demonstrate some of the properties of accelerated motion (whatever the cause of this acceleration may be)— meaning thereby a motion such that the moment of its velocity goes on increasing after departure from rest in simple proportionality to the time. . . .

And he separated the discussion of such " natural acceleration "

---

* Developed in the fourteenth century, and associated with the mathematical philosophers of Oxford, the Rule stated that if the velocity at the start of a uniform acceleration is v, and that at the finish V, the effect (i.e. the distance traversed) is the same as if during the same time the velocity had been of the uniform value $\dfrac{V+v}{2}$, or (if $v=0$), $\dfrac{V}{2}$.

from that of the motion of falling bodies, for (he said) what he aimed to establish was a series of abstract propositions. Only

> if we find that the properties demonstrated later are reproduced by freely falling and accelerated bodies, we may conclude that the definition assumed does include such a motion of falling bodies.[33]

As with the law of inertia and motion in a vacuum, mathematical science is portrayed as idealist ; we can ascertain a correspondence with natural occurrences only as a later step to be undertaken when the abstract theory has been discovered. In fact, verification of Galileo's original concepts by their applicability to the real world could only be carried out when their consequences had been thoroughly explored.

The chief of them was the law of fall already announced in the *Dialogues* of 1632. The proof of this law also utilized the Merton Rule, now formally demonstrated by Galileo with the aid of the " theory of indivisibles " discussed in the First Day. In figure 3, AB represents the time of the accelerated motion ; the infinite instantaneous velocities corresponding to each of the infinite instants between A and B are represented by the little parallel lines drawn across to the hypotenuse AE. Necessarily these increase in the ratio of the times, as Galileo's hypothesis demands. Make K the midpoint in time between A and B, draw IK parallel to AB, and complete the rectangle ABFG. For each part of one of the little parallels exceeding the mean value of velocity IK, there is an equal part of another deficient. Therefore the sum of all the little parallels in the triangle AEB is the same as the sum of all the little parallels in the rectangle ABFG ; or, the sum of the increasing instantaneous velocities in the accelerated motion is the same as the sum of the uniform instantaneous velocities at the mean speed IK. At this point Galileo (like his predecessors in such reasoning) took a great intuitive leap. He declared that hence it followed that the distance traversed, if the velocity increased uniformly from zero to EB, was the same as that traversed in an equal time at the uniform speed IK ;

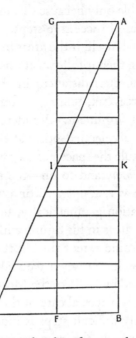

FIG. 3. Natural and uniform acceleration

which implied (but Galileo did not say so) that the distance is represented by the area of the triangle in accelerated motion and the area of the equal rectangle in uniform motion. Although Galileo knew that distance is measured by multiplying time and speed in uniform motion he felt reluctant to express geometrically a distance by an area, because this would have made him say that an infinity of little lines, each representing a velocity, added up to something different, a distance. He lacked a sufficiently clear conception of the infinitesimal calculus to effect such an integration justifiably, though he could perceive what the answer must be. Galileo's lack of an adequate mathematical method for handling continuously varying quantities (such as the velocity of an accelerated body) caused a gap in the logic of his argument now, as in earlier years it had led him into actual mistakes. The great advances in seventeenth-century mechanics were possible only because this mathematical obstacle was surmounted in a series of successive steps.

As for the law of fall itself, if the areas in the figure be accepted as representing distances, then it follows at once that in successive equal time-intervals AK, KB, etc., distances in the ratio 1, 3, 5, 7, · · · represented by the areas AIK, IKBE, . . . must be traversed. Galileo employed a different argument. Rendered more briefly it was : if the velocity of the accelerated body at t is v, and at $2t$ is $2v$ etc., then by the Merton Rule the space passed over after time t is $\frac{1}{2}vt$, after $2t$ is $\frac{1}{2}.4vt$, after $3t$ is $\frac{1}{2}.9vt$ and so on—the generalisation is obvious that the distances are as 1, 4, 9, ... or as the squares of the times. Galileo did this operation geometrically, but it still enabled him to avoid identifying the areas in his figure with the distances traversed.

We can grasp to some extent the effort of mind involved in the attainment of this conclusion when we realise that it is fact implicit in the first result proved by Galileo, the Merton Rule, which becomes in later kinematics only a special case of the law $s = \frac{1}{2}at^2$. What cost Galileo two supremely difficult steps is indeed only one. The nub of the matter is of course the comparison between two pairs of identities (made apparent in modern notation) :

$$(1)\ \frac{ds}{dt} = v\ ; \qquad (2)\ a = \frac{dv}{dt}$$

and $\qquad (1)\ s = vt\ ; \qquad (2)\ at = v$

Galileo recognised only the latter pair, and of that only the first member was perfectly clear to him. He would not have defined either velocity or acceleration as a ratio, though certainly he perceived in principle the implications of such a definition. Finally, he had but clumsy equivalents for obtaining the second pair of expressions as integrals of the first. Galileo had to work with physical concepts that were cloudy and mathematical methods that were cumbersome.

At this point the realist Simplicio rebelled once more : how can we be sure that this fine mathematical formula really applies to falling bodies on Earth ? Salviati replied with a narration of the one

Galilean experiment everyone knows—the rolling of a ball down an inclined plane :

> in such experiments, repeated a full hundred times, we always found that the spaces traversed were to each other as the squares of the times, and this was true for all inclinations of the plane.[34]

However striking, Galileo's positive emphasis was a false move. It was false to the theoretical cast of his science, for it led to the grossly erroneous belief that the validity of the law of fall rested only on such experiments. It was also false to the idealist position clearly maintained in the *Dialogues*. Galileo could indeed demonstrate mathematically that the descent of bodies along inclined planes can be deduced from the law of vertical fall, so that (in theory at least) experiments with such planes could take the place of experiments on free fall. What he could not do was identify the ideal conditions of the theory with those of an actual experiment : as he had said in the *Dialogues*, a ball is never a perfect sphere, nor a piece of timber a perfect plane. When later experimenters (such as Mersenne) rolled balls down planes they did not find the *exact* conformity with theory that Galileo had led them to expect; their experiments would have been the worse if they had. In seeming to ignore the discrepancies attributable to friction and air-resistance, which could in turn be subjected to further mathematical analysis, Galileo seemed to be seeking to geometrise not merely the abstract, Euclidean space of theory, but the physical space of experiments as well. Since the latter can manifestly succeed only when far more complex terms are used, he incurred some danger of jeopardising his whole philosophical position regarding the usefulness of mathematics in physics, so admirably stated in the *Dialogues*. For the moment in his enthusiasm Galileo had retreated to a naive empiricism.

The reason is not hard to find : in this instance naive empiricism taught the same lesson as Galileo's Platonic idealism that nature's language is mathematical. In this respect the *Discourses* are less cautious than the earlier *Dialogues*. The Fourth Day, given over to a wholly original mathematical analysis of the motion of projectiles, shows Galileo at his most confident as a mathematical theorist. He once referred to the discovery of these theorems as that which he cherished most ; it was not merely that he had revealed the power of geometry to comprehend what had for centuries appeared beyond precise comprehension, but that all was unfolded with flawless logic from the basic concepts of motion. Combining the law of inertia with the law of fall Galileo first demonstrated the parabolic trajectory of a projectile (neglecting air-resistance) ; from this could be obtained its height, speed, range and momentum under all conditions. Any fresh question required no more than a fresh recourse to geometry to provide the answer ; it seemed as though physical science could have the unbreakable consistency of mathematics itself. It is not surprising that Galileo was somewhat over-powered by his success, nor that he failed to point clearly to the practical limitations of his new theory. In its beautiful geometric simplicity it seemed that it must be true for the world of experience, and practical men were taken in by Galileo's confidence. The truth proved far more complex.

The other new science of the *Discourses*—the theory of the strength of materials, here begun by Galileo—ran into the same difficulty. Galileo's insight was admirable. He showed clearly why a model is relatively stronger than the full-scale structure of scale dimensions ; why the bones of a giant would have to be dispro-portionately thick for their length. By introducing the notion of particulate or atomic composition he suggested that part of the cohesive strength of solids might derive from the effect of *fuga vacui*. He considered the varied tensile strengths of different materials ; the optimum shape for an arch ; the optimum longi-tudinal profile of beams, and so on. For the first time a serious effort was made to create by scientific analysis a body of theory for civil engineering and building practice. Most of his statements in relative terms are correct, but Galileo's theory of the absolute strength of beams was spoiled by one error of concept : the neutral axis, or pivot about which a beam breaks, is well inside the beam and not at the base of the fracture. There is an effect of compression that Galileo did not foresee.

Inevitably in such matters of applied mechanics, as in the application of mechanical principles to the revolutions of the heavens, Galileo left much for his successors to do. It was their task both to define the principles with greater exactitude and completeness, and to show (no less mathematically) how the principles actually operated when the real complexities of the physical world were more fully examined than they were by Galileo himself. Neither the logic of the mathematical physics he inaugurated nor its agreement with observable phenomena were quite adequate until long after Galileo's own death. It is important to recognise this, not in order to lessen Galileo's achievements—which are too great to need a perfection they did not attain—but in order to understand what he accomplished, and what problems he left unsolved, sometimes unperceived. Galileo was almost always intuitively correct in his new ideas and his perception of the physical world was not only original throughout but thoroughly sound. Yet he was too much engaged with his own emancipation from the older science to express his own new perception with sharp definition and entire rigour. The intellectual struggle from which Galileo emerged as the first modern physical scientist left cloudy traces on his concepts and his reasoning alike. They were rough-hewn and roughly cemented together.

Yet, after all, the impression made by Galileo's two great treatises upon the science of his century—and all subsequent science —derives not from the imperfections of his new mathematical and mechanical view of the universe but from the tremendous power of this view to explain things. What most aroused the admiration of Galileo's contemporaries was his success in creating a mathematical science of motion and, closely related to this, his persuasive justification of the Copernican theory in which again Galileo's own discoveries with the telescope played a notable part. Nor did they miss the innumerable other interesting observations and novel ideas of the two treatises, most of them illustrating Galileo's philosophy that true understanding of natural phenomena assumes the form of a mathematical theory : the discussion of musical harmony in the *Discourses*, for example. Galileo's influence was much wider than that of his great discoveries. His opinions were vigorous and their expression forthright. No reader could fail to absorb his admiration of Archimedes and contempt not so much for Aristotle himself as for the unworthy Aristotelians who ran to the authority of books rather than facts and logic. Scientific judgements, said Salviati, are no more to be swayed by authority than by rhetoric :

> If what we are discussing were a point of law or of the humanities, in which neither true nor false exists, one might trust in subtlety of mind and readiness of tongue and in the greater experience of the writers, and expect him who excelled in those things to make his reasoning more plausible, and one might judge it to be the best. But in the natural sciences, whose conclusions are true and necessary and have nothing to do with human will, one must take care not to place oneself in the defence of error ; for here a thousand Demosthenes and a thousand Aristotles would be left in the lurch by every mediocre wit who happened to hit upon the truth for himself.[35]

When Aristotle was commended by Galileo it was usually for sound features of his method as when, despite the fallacy of his *a priori* arguments, Aristotle "preferred sensible experience to any argument".[36] His conclusions were rarely approved by Galileo, who made Sagredo declare :

> How many propositions have I noted in Aristotle . . . that are not only wrong, but wrong in such a way that their diametrical opposites are true ![37]

Quite apart from such sweeping rejections of tradition, the way in which the old paraphernalia of physical science was jettisoned without regret—crystalline spheres, forms and qualities, natural and

74

violent motion, sympathies and antipathies, the impossibility of a vacuum and all the rest—was enough to make an Aristotelian like Simplicio plead for mercy. Bewildered in a world of ideas whence all familiar landmarks had been removed, where reason was now proved to be unreason, where even the customary language of natural philosophy was thrown away as gibberish, there was only one alternative to the surrender that Galileo, for all his formal deference, demanded.

Neither softening his phrases nor overlooking an absurdity he assailed his opponents' ignorance and his readers' prejudices with equal fervour. He spared them nothing, for this was no idle debate. In 1632 Galileo wrote to circumvent the suppression of the intellectual revolution of his time that he knew was impending. He was a leader of that revolution and he believed that its suppression would bring both science and religion into disrepute. In 1638 he published his scientific testament from a domestic prison where he was denied professional conversation save under the strictest supervision. In both books he had all cause to use every weapon. He accused the Aristotelians of not understanding the words of their master; of being ignorant of both the Ptolemaic and the Copernican geometries; of being blind to the truth before their noses; of building philosophies out of mere words. At times he would quibble to score a point against them. He was not often unfair but he kept up the pressure remorselessly. Yet only rarely did Galileo claim that the Aristotelians erred in their facts because they had not experimented to discover them. Rather it was their failure to consider facts correctly that he constantly stressed. His chief target was the absurdity of their reasoning. Why did Galileo devote several pages to an obscure critic's fantastic argument about a cannon-ball falling from the moon? Because it held the childish mistake of making the radius of a circle six times the circumference, instead of less than one-sixth, and revealed ignorance of the elementary fact that according to Copernicus the diurnal and annual revolutions of the Earth take place in the same direction. Galileo did not seek to overwhelm hostility by a wealth of new evidence—he could not ; he sought to make thought, like the Earth, turn into the light. As Simplicio well put it under Salviati's tuition, " I perceive the darkness passing away from my mind ".

It was an intellectual revolution that Galileo effected. Jupiter's satellites, Venus' phases, the swings of a pendulum were convenient and indeed essential aids to the imagination; they helped to render the new ideas more concrete and to peg the new physics to reality. But they were not instrumental to the new physics. To Galileo the senses were always deceptive ; as he had described in a famous passage of *The Assayer* they portray a world of illusion, not the physical reality of shape, number and motion.[38] The entrance to this real world was through reason. Like Aristarchos and Copernicus Galileo had at one time noted the difficulties confronting the heliocentric theory without being able to resolve them. He too had been confident of what his reason told him must be so, in spite of solid appearance to the contrary. In Salviati's words :

the experiences which overtly contradict the annual movement [of the Earth] are indeed so much greater in their apparent force [than those against the diurnal rotation] that I repeat there is no limit to my astonishment when I reflect that Aristarchos and Copernicus were able to make reason so conquer sense that, in defiance of the latter, the former became mistress of their belief.[39]

As with astronomy, so with the science of motion reason must be the mistress of belief. Galileo's aim was to show that his definitions and theorems made experience coherent, just as the Copernican system had made the heavenly revolutions coherent. He could no more prove the former than the latter. In fact when, outstripping the bounds he had set himself, Galileo produced in the Fourth Day of the *Dialogues* of 1632 the argument that he thought turned the tables, in a theory tracing tidal motion to the rotation of the Earth, he gave expression to the worst of all his intuitions. His attempt to prove the Earth's motion *de facto* was the weakest thing he ever did. For all its superficial plausibility (though it made the tidal surge occur once a day instead of twice) it conflicted with Galileo's earlier, correct enunciation of the properties of inertial systems. It led him into an unfortunate " respectful reproach " of Kepler who had " lent his ear and his assent to the moon's dominion over the waters, to occult properties, and such puerilities ".[40] Over-eagerness had brought him into the trap he had so often sprung upon his enemies. In truth Galileo was still partly with Copernicus: just as he could not quite perfect the science of motion, he could not yet quite complete the alliance between reason, teaching him how things must be, and empirical inquiry teaching how they really are.

CHAPTER III

NATURE'S LANGUAGE

SIMPLICIO : Believe me, if I were again beginning my studies, I should follow the advice of Plato and start with mathematics, a science which proceeds very cautiously and admits nothing as established until it has been rigorously demonstrated.[1]

There has always been a link between science and mathematics. From its very beginning science has included a metrical element of varying importance : whether in charting the positions of the stars and timing the revolutions of the planets, or in more mundane endeavours, whenever any argument turned upon the relative numbers or sizes of things a rudimentary mathematical operation was involved. In astronomy, especially, complicated methods of calculation early came into use. Since mathematics itself is concerned only with relations between numbers (or, in geometry, lines representing numbers) its theorems have no necessary relevance to the physical world that science describes and explains, but in a few respects the correspondence between mathematical theorems and physical reality—especially the geometrical properties of actual bodies—has always been so obvious that it could not be overlooked. Doubts only arose when the attempt was made to press this correspondence beyond the obvious, in order to infer that reality has concealed mathematical properties that are part of its essence. No deep sophistication is required to treat fields as rectangles and triangles, but Pythagoras' discovery of the relations between the lengths of taut strings and the notes they yield when struck marks a harder step. For here the mathematical ratios easily measurable in the strings themselves were, by extension, ascribed to the sounds which were not directly measurable, at a time when the concept of measuring sound did not even exist. The mathematisation of nature advanced yet further when mathematical properties that could not be verified at all appeared in physics, as they did in Plato's *Timaeus*.

Such transitions may seem to be of degree only. In principle there is, perhaps, no greater artificiality in supposing that the ultimate particles of matter are perfect spheres than in treating a farmer's field as though it were a perfect square. Yet it seems reasonable to make a distinction between those applications of mathematics to physical questions where directly measurable quantities alone are used, and those in which unmeasurable quantities enter ; and between that use of mathematics which provides a result (a number or magnitude) and that which offers a proof that some idea is or is not plausible. In the latter case mathematics is used purely as a logical device, to show that if the world is like this in one respect, it must be like that in another. Such a logical device teaches only the lesson of ˙consistency, it does not teach how the world is.

There are, therefore, two versions of the same question that may be asked about the usefulness of mathematics to physical science. It may be asked : Are the relationships between things always expressible in mathematical terms ? or, Is the test of consistency one to which ideas about nature must always be subject ? Plato would have answered yes to both questions, Aristotle no. The Middle Ages moved from a confidently Aristotelian to a doubtfully Platonic position, while the seventeenth century adopted a modified Platonism. It was not, of course, a question of believing that mathematical reasoning only was suitable for science, but rather that this form has greater precision and rigour than verbal argument. As against the Aristotelian position that " in physical matters one need not always require a mathematical demonstration " the Platonist simply retorted, " Granted, where none is to be had, but when there is one at hand why do you not wish to use it ? "[2] One learns from Galileo what this means : the space in which planets revolve and apples fall is Euclidean space and has all its properties ; the motions that bodies actually have are those described in kinematic theory, itself a product of geometry. At least in these simple respects the universe must be conceived to be as logical and consistent as Euclid's theorems, and God becomes the first geometer. In other words, mathematics is not just an abstract science exploring the relations of numbers, for in these relations lies a model of physical reality.

Archimedes, whom Galileo never named without praise, had pointed the way to mathematical physics. He had shown how basic physical properties could be so defined in statics that their negation was inconceivable (though not impossible), and how their consequences could be explored by geometrical reasoning. The fourteenth century had further investigated types of mathematical relationship appropriate to physics, but not ones that could clearly be shown to hold in the physical world. Thus some elements of the seventeenth century's application of mathematics to natural science were already old—and indeed in optics and astronomy these older elements were conspicuous. But the true Archimedean type must be distinguished from weaker attempts to mathematise nature. Leonardo's studies of proportion in anatomy, like those of Renaissance architects, imply a superficial Platonism rather than a profound vision—such as Galileo introduced into physical science—that the theory and explanation of natural phenomena are to be couched in mathematical form. Similarly Tartaglia, attempting in the generation after Leonardo to portray the trajectory of a projectile as a geometrical curve, had done so without a mathematical theory. Perhaps the purest form of Platonism in science, however, and that farthest removed from Archimedean mathematical physics, attributed geometrical characteristics to natural phenomena on metaphysical grounds alone. It is exemplified in Galileo's own attachment to the ancient belief that celestial motions must be perfectly circular. He could envisage no reason why bodies in free, eternal movement should follow any less complete and unvarying line. No doubt this idea had rational roots in observation, as much as Euclid's postulate that the path of light is a straight line, but even in Plato's day it was known that simple concentric spheres would not suffice to carry round the planets ; in Galileo's the spheres had gone altogether. By contrast the investigations of Kepler, who found strength to renounce the ancient prejudice, reveal the true Archimedean spirit. The three laws of planetary motion, laboriously extracted from the observational data, stand parallel to Archimedes' definition of the conditions of static equilibrium though they are the less self-evident as the phenomena they describe are the more complex. As Archimedes' axioms are the foundation of the mathematical science of statics, so Kepler's laws were to become the foundation of the mathematical science of celestial mechanics.

Kepler solved the problem of the planetary orbits not by his tedious method of plotting positions and endeavouring to draw a line through them, but by suddenly seeing that the ellipse—which he had already employed as a calculating aid—could be so applied that it satisfied all his needs. It was a regular, well-known curve and he at once concluded that if the orbit of Mars was an ellipse, all the orbits must be elliptical. Galileo in like fashion found that the trajectory of a projectile was a parabola, another familiar conic section. In the same period Willebrord Snel (1591–1626), experimenting on the refraction of light, discovered the trigonometrical ratio or sine law of refraction known by his name. Other eponymous relationships between physical quantities were rapidly discovered—Boyle's Law, Hooke's Law, Newton's law of gravitation—all having a simple mathematical form. Whether deduced like Boyle's Law from empirical results or not, each relationship has in it the element of idealism occurring in Kepler's discovery of the planetary ellipse. No one could have proved that gravity decreases as the second power of the distance exactly, and not as the

1.999...th power, nor that Snel's law holds for every discoverable refracting material (in fact it does not). The universality of such exact mathematical relationships as the seventeenth century began to call laws depends in part on a Platonic faith, a mathematical version of the assumption that Newton expressed as "Nature is always simple and conformable to herself". One can hardly fail to detect it in examples of mathematical theorisation that proved to be incorrect. If Galileo was the more inclined to believe that trajectories must be parabolical because nature works in geometrical ways, the same influence of a Platonic outlook may have been stronger still when he assigned to circular motion physical properties, such as isochronism, which he could not rigorously prove. Though he erred in this, at least his mathematical philosophy was justified by Huygens' discovery that motion in another simple geometrical curve, the cycloid, is truly isochronal.

Yet far more important for science, in the long run, than the mathematical idealism of Plato was the mathematical logic of Archimedes, from which it followed that once a determinate cause was established in physical theory, the effects flowing from it were mathematically deducible. Assuming no defect in the mathematical reasoning, the detection of these effects in nature constituted a proof of the correctness of the determination of the cause. Galileo's gigantic achievement was to extend this method to the science of motion, giving it a mathematical rigour, and in so doing establishing the universal validity of such a science. As he had modelled himself upon Archimedes, so later physicists modelled mathematical theory upon Galileo. And his was by far the more necessary and instructive example, because his achievement was both conceptual and mathematical. Conceptual, in arriving at definitions of the physical quantities inertia and acceleration; mathematical in arriving at the method of geometrical manipulation. Archimedes had won a very limited territory; Galileo's was potentially unlimited because mathematical physics had forced an entrance into the world of things that move and change. The victory of the new mathematical physics did not only set the Earth free to spin, it opened the possibility for a fresh vision of the cosmos as an organised pattern of mathematically regulated motions. It offered the essential substitute for the statically ordered cosmos of Aristotle.

The spirit of Archimedes did not triumph over the shade of Aristotle without a struggle. The opinion that physical bodies do not conform to geometrical canons of uniformity, circularity, planeness and so on tended to suggest that mathematical reasoning could only be really true for conditions of impossible simplicity. An analogous objection was that mathematics only " saved the phenomena " without reflecting the real nature of things. Ptolemy had written of the eccentrics, deferents and epicycles of Greek astronomy as mere hypotheses and the same view had been extended to Copernicus' *De Revolutionibus* in Osiander's preface to the book. Rome itself had countenanced the heliocentric hypothesis as a mere calculating device ; if the object of mathematical astronomy were to discover when Saturn and Venus would next come into conjunction (for example) it did not matter much what kind of procedures were used to grind out the answer. The opposition of the Roman Church was to the idea that such a mathematical system as the Copernican did represent the physical structure of the universe. It seemed—plausibly enough in the state of astronomy then—that mathematicians could devise endless ways of portraying the same motions, none of them necessarily true, so that a mathematical argument could hardly deserve the name of a scientific theory. Yet if mathematicians could be regarded as mere drudges, they could also be accused of claiming a certainty of knowledge belonging only to God. It would be blasphemous (on this view) to suppose that the world is mathematically constructed, and that man could fully comprehend its mathematical principles, for this would make man's knowledge equal to the divine. Galileo indeed declared boldly that while man knows only a few of the mathematical truths perceived by God, his understanding of these few " equals the divine in objective certainty, for here it succeeds in understanding necessarily, beyond which there can be no greater sureness ".[3]

The point to which Galileo was replying—not quite adequately—is a far from trivial one. How can the nature of mathematics as a formal science, whose propositions are necessarily true, square with that of physical science, where propositions are only contingently true ? In Euclidean geometry the theorem known by Pythagoras' name cannot ever be falsified, but there is no statement in natural science that may not be modified in the light of subsequent discoveries. An answer may be found in the contingency of both the physical laws on which mathematical reasoning rests, and of the experimental tests by which it is verified. Elementary geometrical optics does not apply in cases where light is not transmitted in straight lines. But it is not surprising that the Platonic idealism of the seventeenth century prevented Galileo—or for that matter Newton—from apprehending the limitations of mathematical certainty in science in clear terms.

Galileo dealt with the difficulty by asserting a metaphysical principle :

Philosophy is written in this grand book, the universe, which stands continually open to our gaze. But the book cannot be understood unless one first learns to comprehend the language and read the letters in which it is composed. It is written in the language of mathematics, and its characters are triangles, circles, and other geometric figures without which it is humanly impossible to understand a single word of it ; without these, one wanders about in a dark labyrinth.[4]

This was no empty view that harmony, proportion and geometrical form occur everywhere in nature. It was justified by results, by the correspondence of theory with observation that it could ensure. In both the *Discourses* and the *Dialogues* this view furnished the metaphysical basis for theories that explained the behaviour of bodies in motion from the definitions of motion. Mathematics was shown to be the vehicle of scientific explanation. Galileo showed how to extract from the bewildering variety of physical experience mathematically definite concepts, and how—through mathematical reasoning—general conclusions followed. Using modern notation, if velocity is defined as $\frac{ds}{dt}$ and acceleration as $\frac{dv}{dt}$, the conclusion $s = \frac{1}{2}at^2$ cannot be escaped ; and it is one that can be verified experimentally. Adopting Galileo's method the seventeenth century substituted for *weight* (the sense of muscular strain in lifting bodies) the mathematical concept mg ; for *momentum* (the sense of muscular strain in arresting anything that moves) $\frac{1}{2}mv^2$; for *force* (the sense of muscular effort in moving something) ma, and so on. It created quantities directly susceptible of the operations of geometry and algebra and went on to explore the consequences. It might seem that this was simply a question of definition, and so it was ; but not of descriptive definition. No one ever saw a force as mass multiplied by acceleration, in the way that a table may be seen as a plane on four legs. In physical science on Galileo's model the concept and the definition are the same thing ; the concept of weight is mg and cannot be anything else. The expression mg is not a mathematically exact form of some idea of weight that had existed before ; it is a new concept.

Galileo created a pattern of science of tremendous power. Even in the mid-twentieth century by no means all science falls into this pattern, but it has been extended over the whole of physical science and has penetrated biology. If its immediate conspicuous success— great enough, one must allow, since it transformed man's idea of the universe—was confined to the science of motion, there were already hints in the *Discourses* of the manner in which mathematical language could be made to speak of other subjects. To Galileo it was clear that, as the varied phenomena of nature arise from the motions of its component parts, their explanation must ultimately assume a mathematical dress. Hence the prospect appeared not only of mathematising the single branches of science but of bringing them together in a unified mathematical theory. If the book of nature employed none but mathematical characters there could be no unsurmountable divisions between its chapters. This is the view of nature that physical science between the time of Galileo and that of Newton sought to elucidate, and that Newton himself established in the *Mathematical Principles of Natural Philosophy* (1687).

It is obvious that for science to develop mathematical theories its ideas must be expressed in quantitative terms. To gain a clear notion of velocity as a ratio it is necessary first to understand what is meant by the measurement of time and distance, though it is immaterial from the theoretical point of view whether either can be measured very accurately. Whatever is expressed by a or m or t algebraically, or by a line or area geometrically, must have potential quantitative significance or the procedure is without meaning. At the experimental level this increases the importance of instrumental measurement. Certainly the idea of a scale of heat, say from 0–10 degrees of heat, is older than any thermometer other than the human body and its perception of gradations in heat. The concept that variations in a quality of a thing, like its heat, could be rendered in numerical terms was itself an important step towards quantitative physics. Yet there is little chance of using the concept effectively till some knowledge of the working of the scale in nature can be gained by means of a thermometer that gives numerical readings. So also pressure, easily conceived statically as weight divided by area of contact, is a concept hard to extend to gases without the aid of a barometer. The reverse effect of the introduction of mathematics into science is no less striking. As M. Koyré has pointed out, the qualitative science of Aristotle resisted mathematisation because whatever is in principle thought incapable of yielding quantitative concepts must disappear from mathematical science as being without significance. On Galileo's position whatever is not mathematical in nature is not really there at all ; it is merely a sensation, a response of the human nervous system—or the imagination—to stimuli which are mathematical. The sense of pleasure in musical harmony is in man, not in nature, which recognises only the

mathematical periodicity of sound-vibrations ; the sense of warmth
—pleasant or hurtful—is purely subjective, the physical reality being
the minute motion of invisible particles. In fact, the mathematical
idealism of Plato combined in Galileo and his successors with the
particulate theory of matter deriving from Democritos and
Epicuros to rule out the qualitative physics of Aristotle. No more
powerful union of ideas has ever occurred in the history of science.

Once the theory of a branch of science has been given a mathematical structure there are only two limitations to its development:
that of the ability to frame adequate concepts, and that of the ability
to surmount purely mathematical difficulties. (Experimental
problems are not insignificant, but they can be considered as an
aspect of the problem of formulating suitable concepts.) Questions
of mathematical physics can obviously only be solved by appropriate
mathematical techniques ; should these not exist, the advance of
physics is obstructed. Without knowledge of the mathematics of
conic sections it would have been impossible to resolve many
problems of optics, for instance. The more complex discovery of
Kepler's laws is very instructive, for if Kepler had not known the
geometry of the ellipse from conics he would have been forced to
resort to combinations of circular and rectilinear motions like his
predecessors. These might have been accurate enough for purposes
of prediction, but they could not have had the physical significance
of the ellipse—that the sun occupies one focus of each planetary
orbit. From the mathematical point of view one representation
may be made equivalent to another (to any assigned limits) ; it
does not follow, however, that the physical usefulness of the two
representations will be equivalent. The greater the resourcefulness
of mathematics, the more simple and comprehensive its expressions,
the richer will be the physical potentialities.

The progress of mathematical science in the seventeenth century
and the crowning achievement of Newton would never have
occurred if mathematics itself had not made giant strides. In the
time of Galileo mathematicians' resources were already stretched

almost to breaking-point by the practical problems with which they contended—even in such an apparently trivial matter (which Kepler tackled) as calculating the volume of a wine-cask. Astronomy was notorious for the tedious computations it imposed. These were greatly facilitated by the use of logarithms, of which the first full table (numbers 1 to 100,000) calculated by Henry Briggs (1561–1630) and Adrien Vlacq (c. 1600–1667) was published in 1628.[5] The nature of logarithms was not clearly understood at first, but their usefulness in other ways than simple computing appeared later.[6] Since multiplication is effected by adding two logarithms and division by subtracting them, it was practicable to represent logarithms on a pair of scales so that these operations could be effected mechanically and the answer read off directly. This step was taken by the English mathematician William Oughtred (1575–1660) in 1622, but the slide-rule was not much used in the seventeenth century, being too crude for astronomers and too refined for engineers. The same fate attended the calculating machines devised by Pascal (1642), Leibniz (1671) and Morland (1673).

Of far greater interest to mathematicians were the problems that arose in the theory of physics. A simple example occurs in optics.

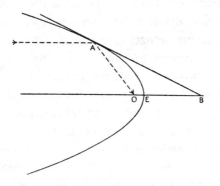

FIG. 4. Reflection from a paraboloid

Suppose a concave mirror, formed by rotating such a curve as AE about the axis OE, is required to reflect all incident parallel rays of light to the point O. Clearly, since each incident ray is reflected from any point A at the same angle as that of the tangent to the axis, AO must be equal to OB for any position of the tangent AB. This is a physical requirement but it is a mathematical problem to discover the curve that will satisfy it. Kepler proved (1604) that the parabola has this property, so that a burning-mirror should be a paraboloid. (The same argument applies to the reception of radio-waves of very high frequency.) Here the physical principle—equality of the angle of incidence and reflection at the surface of the mirror—is a simple one; for lenses the similar problem of determining the precise shape for accurate focus is more complex because the sine-law of refraction is less simple, and two surfaces have to be considered. Descartes explored this problem in 1637, and many subsequent attempts were made to solve it. A general solution for thick lenses was given by Edmond Halley (1656–1742) in 1693.

The mathematical interest of these problems is that they turn on the theory of tangents, a central interest of seventeenth century mathematics. The heart of the matter was : given a point such as A on a curve (as in the last figure), where would the tangent at A meet some other defined line, such as the axis OE ? For if two points A and B are defined in this way, the tangent is defined. For some particular curves, like the parabola, solutions could be found by the geometrical methods of the early seventeenth century. But no general solution was possible, because there was no general way of expressing the characteristics of a curved line. Curves could only be defined by the method of their production : thus Archimedes' spiral was defined as the curve traced by a point moving uniformly along a straight line while the line uniformly rotates about one of its ends. How restricting this manner of definition was may be gathered from Galileo's inability to define the curve (catenary) which a slack rope hung by its ends assumes. There was no language or notation in which Galileo could begin to describe this physical form mathematically, and he could only guess that it was close to parabolical. Yet Galileo was well aware that any mathematical curve is the product of a pattern or regularity, not arbitrary like a snail's trail.

A related problem—though the relation was only perceived about mid-century—was that of calculating areas or volumes. Again, some special cases—the circle and sphere, the parabola and paraboloid—had been solved in antiquity, but for the same reasons as before no general methods had been attained. Such calculations were of scientific and even commercial importance, as well as of high intrinsic scientific interest. The most fruitful approach of Galileo's time was through the assumption that a line could be regarded as being composed of an infinite number of adjacent points, a surface of an infinite number of parallel straight lines, and a volume of infinite superposed surfaces. In the *Discourses* Galileo maintained that a line (and the same applied to areas and volumes) is a " continuous quantity built up of an infinite number of indivisibles ". For, on the one hand, if the number of parts were finite, division of the line could not (mentally) be continued indefinitely; and, on the other hand, if the parts themselves were finite, or divisible, the line itself would become infinite. Hence (Galileo argued) a finite quantity arose from the multiplication of the infinitely large by the infinitely small, an idea fully according with Galileo's atomistic philosophy.[7] His ideas owed something to Kepler, and to his mathematical friend Luca Valerio. In turn Galileo's teaching strongly influenced Bonaventura Cavalieri, the author of the foremost treatise of the time on the geometry of indivisibles (*Geometria indivisibilibus*, 1635).[8] Cavalieri showed the summation of lines and planes by means of series. The right triangle is an easy case, since the lines composing it have lengths $0, 1, 2, 3, \ldots n$, of which the sum is $\frac{1}{2}n^2$; since by the same argument, all the equal lines in a rectangle of the same base add up to n^2, the triangle's area is half the rectangle's. In a cone, infinite circles of area increasing as $0, 1, 4, 9, \ldots n^2$ yield the sum $\frac{1}{3}n^3$ as n approaches infinity, and since the equal circles of an enclosing cylinder add up to n^3, the cone has one-third the volume of the cylinder. Confirming his reasoning—which depended on the possibility of summing infinite series—by these well-known results, Cavalieri went on to express the new general relation, $\dfrac{1^m + 2^m + 3^m + \ldots n^m}{n^m + 1}$ approximates to $\dfrac{1}{m+1}$ as n approaches infinity, m being a positive integer. Later (1656), John Wallis extended this theorem to the cases where m is not an integer.

Despite the modern notation just used for simplicity Cavalieri was a geometer, not an algebraist. A feature of his work that seems evident to posterity, the alliance of algebra and geometry in his application of series, was not clearly brought out. Nor was Cavalieri able to define the nature of indivisibles satisfactorily ; if lines and planes have no thickness, how can they be added together to form a surface or a volume ? How does the infinity of points in one line compare with the infinity of points in one twice as long ? Cavalieri did not clarify the mathematical concept of infinity, of which Galileo had said that it transcended understanding, for he was little interested in the criteria of logical rigour in mathematical reasoning. Consequently his methods were distrusted by many contemporaries. It is curious—and hardly coincidental—that at the very time when the idea of the infinity of the universe was entering cosmology, it was also penetrating mathematics. In both contexts the idea of infinity, rejected by the Greeks, now seemed necessary to the imagination and yet to baffle comprehension. Precisely because the early concepts of mathematical infinity were so closely related to physical ideas, they could not illuminate the concept of infinite space.[9] The formulation of a correct notion of infinity in physical science had necessarily to follow upon a clarification of the idea of infinity in pure numbers; and that was scarcely accomplished in the seventeenth century.

The alliance of geometry and algebra was solidly welded in a work published two years after Cavalieri's, La Géométrie of René

Descartes annexed to his *Discours de la Méthode pour bien conduire sa raison et chercher la vérité dans les sciences* (1637). Analytical geometry had indeed been employed earlier—but not published—by Pierre Fermat (1601–65), and the use of co-ordinates to determine a position (as on a map) or to define a motion (as in the Merton Rule and Galileo's demonstrations) was older still. But it was from Descartes that the new system was learnt. The key to analytical geometry is deceptively obvious :

> All geometrical problems may easily be reduced to such terms that afterwards one only needs to know the lengths of certain straight lines in order to construct them. As arithmetic consists of four or five operations only (namely addition, subtraction, multiplication, division and the extraction of roots, which is a kind of division), so in geometry to find the lines required it is only necessary to add or subtract others; or, given two lines, to take a third line as unity and discover a fourth proportional (which is multiplication) . . .[10]

Designating the lines in a geometrical figure by letters and forming equations between them, the solution of the equations solves the problem by giving the length of an unknown line. For Descartes such an expression as $(a + b)^2$—the square of the sum of two lines— was itself a line, and not as in geometry an area. The transposition into algebraic symbolism was facilitated by representing all lines in the figure in terms of x and y, two co-ordinates normally rectangular. Map-co-ordinates define such a position as P statically. Descartes designated the point P kinematically, so that as P assumes every position possible from the equation defining it, it traces a straight line or curve. The equations $y^2 = r^2 - x^2$, $y^2 = 4ax$, $y = ax + b$ respectively define such motions of P as will trace a circle, a parabola and a straight line; not only could any equation of terms including x and y be represented geometrically in this way, but classes of curve were found to correspond to classes of equation.

Analytical geometry rendered possible the later achievements of seventeenth-century mathematical physics ; without this method the application of mathematics to science would have been stultified. The problems of physics—especially mechanics—presented themselves in spatial terms, that is, geometrically ; by Descartes' discovery they could be subjected to the flexible and solvent attack of algebra. To take a simple instance, as the trajectory of a projectile (neglecting air-resistance) is shaped by two variables only, the angle and velocity of projection, Descartes' analytical geometry denoted the parabola traced by the equation

$$y = x . \tan a - \frac{gx^2}{2V^2\cos^2 a}$$

Assuming the surface to be level, when the projectile touches it the height (y) is zero, from which it easily follows that the range is $\frac{V^2 . \sin 2a}{g}$. Sin $2a$ is a maximum when $a = 45°$, which is therefore the angle of projection giving maximum range at a given velocity. Because the trajectory is an upright parabola, the projectile is at its greatest height when at half-range (when $x = \frac{V^2 \sin 2a}{2g}$), so that from the first equation this height is $\frac{V^2 \sin^2 a}{2g}$. Again, because the horizontal and vertical components of velocity at any point (x, y) on the trajectory are easily discovered, the angle of motion at that point is found to be $\tan \theta = \frac{2y_1}{x_1} - \tan a$. This result shows how a tangent may be drawn to a parabola at any point (for its angle of slope is now known)—an example of the way in which a theorem in pure mathematics may be reached through a mathematical-physical argument. Other properties of the parabolic trajectory, far less obvious in geometry, could be elicited after the same manner ; and it may be noted too that in algebraic expressions each property can be calculated separately whereas geometry would state proportions, such as " the extreme height at any angle is a quarter of the range multiplied by the tangent of the angle ", a more pictorial statement but one less convenient for calculation.

In the two decades after the publication of *La Géométrie* algebra assumed a notation that is still customary. It was still usual to write *aa* for a^2, and a few other variants from modern usage, but essentially the rapid improvement in symbols that had occurred in the last half-century was complete. Despite all the advantages of algebraic expression over geometric, however, the latter was by no means quickly displaced from scientific mathematics. For this there were many reasons. Geometry was still more widely taught than algebra and, more important, geometrical proofs were still regarded as more evident and certain than algebraic ones. Most of the scientific problems that could be handled mathematically posed themselves naturally in geometrical terms, and though their solution might then be best approached by Cartesian analysis, it was in general possible (and still held to be desirable) to construct a synthetic, geometrical proof of the solution. Following the Greek example, analysis in mathematics was considered rather as a means of arriving at a result, than as a rigorous demonstration of its correctness. Only about the close of the century did mathematical physics turn, very rapidly, to algebraic notation. It could no longer be escaped, for by this time the differential and integral calculus was becoming indispensable to physics.

The differential and integral calculus of Leibniz—and the equivalent, earlier method of fluxions devised by Newton—were a fusion of mathematical advances that at first had appeared unrelated, in the general theory of equations, of tangents, of series and of calculating areas. This fusion almost at one stroke furnished an all-embracing solution of the problems with which seventeenth-century mathematicians had contended. What had formerly been very difficult to achieve now became relatively easy. And, just as the calculus furnished the language in which (from the early eighteenth century) the science of motion was most perfectly expressed, so at the time of its development a major stimulus was given by the need to treat mathematically quantities that vary, quantities expressed by curves on Cartesian co-ordinates. Some of the most dramatic incidents of seventeenth-century mathematics related to these mechanically determined problems, notably the challenge issued by Blaise Pascal (under the pseudonym Dettonville) to the mathematicians of Europe, to determine the length of a cycloidal arc. Galileo had made the science of motion mathematical; it became immediately a high test of mathematical ability to solve the questions that this new science presented. For the first time since trigonometry arose to satisfy the needs of astronomers the demands of science really exercised mathematical skill, and the advance of scientific theory went hand-in-hand with that of mathematics.

The application of a tangent to a parabola by a kinematic procedure (employed by Roberval about 1634, but not printed till much later) indicates the way in which consideration of motion could form fertile mathematical ideas. The idea of motion, which was geometrical, provided a transition to the algebraic idea of variation. Thus Napier had expressed his notion of logarithms by comparing the motion of a point whose velocity is constantly decreasing, with that of another point having a uniform velocity. The distances traversed by the former were related to numbers, those traversed by the latter to logarithms. For Descartes curves were paths traced by moving points. The Italian mathematician Evangelista Torricelli (1608–47), in unpublished papers of about 1645–6, also used kinematic methods though not in a generalised, analytical manner. Always in such reasoning the notion of time was implied ; Newton, continuing the tradition, made it specific. Newton considered two points A and B moving along a line with different speeds, which determine the lengths of the two lines, OA, OB (figure 5). At any given moment, whatever the lengths of OA and OB, during an infinitesimally short time the change in length of OA will be proportional to the velocity of A at that moment, and similarly with OB. One can put this the other way round and say that the relative velocities of A and B, which Newton called their fluxions and denoted by such symbols as $\dot{x}$, $\dot{y}$, $\dot{z}$, can be found from

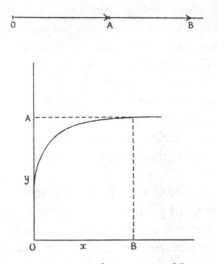

FIG. 5. Geometrical expression of fluxions

the relative changes in lengths of OA and OB. If these two lines are set at right angles on Cartesian co-ordinates then the velocity of increase of y compared with that of x determines the shape of the curve traced, which itself corresponds to an algebraic function. For though the velocities of increase change, the relative velocities or fluxions do not, and can be found from the equation describing the curve.

Here, as in the other approach to the calculus through indivisibles, the idea of limit appears as an essential concept of the calculus. For suppose $y = x^2$. In a very short time interval the increase of y to the increase of x is as $\dot{y}$ to $\dot{x}$. From the function postulated,

$$(y + \dot{y}) = (x + \dot{x})^2 = x^2 + 2x\dot{x} + \dot{x}^2,$$

and subtracting $y = x^2$ from each side

$$\dot{y} = 2x\dot{x}.$$

The very small quantity $\dot{x}^2$ is a dimension smaller than $\dot{x}$, which itself is not (as Newton insisted) a very small finite quantity, but only the "just nascent beginning of a finite quantity"; it is the limit of a quantity as it approaches zero but is not zero. $\dot{x}^2$ can therefore quite reasonably be neglected. Even at the end of the seventeenth century, as Newton's language suggests, this concept was so unclearly framed that it was possible to maintain that though the results given by the calculus were undeniable, they were without logical foundation. This criticism was trenchantly expounded by the philosopher Berkeley in *The Analyst* (1734). Thus Newton, in some of his statements concerning fluxions, had considered them as the " prime and ultimate ratios " of " evanescent " quantities. Berkeley insisted that it was illogical to consider fluxions as products of the ratio $\frac{0}{0}$, and unrigorous to argue that quantities as small as $\dot{x}^2$ were negligible.

Earlier steps towards the notion of limit taken by such mathematicians as Cavalieri, Roberval (1602–75) and Wallis (1616–1705) were even less capable of justification than the fluxions of Newton or the differentials of Leibniz. In Roberval's work, for instance, the idea of limit was confused by that of the indivisible. He proved the well-known relation that a parabolic segment has one-third the area of the enclosing rectangle by summing the areas of separate columns (shaded in the figure). This gave a series:

$$\frac{1+4+9\ldots+(n-1)^2}{n^3}$$

which he took to be equivalent to $\frac{1}{3}$ if n was sufficiently large, assuming that as the columns approached the limit of narrowness the little triangular spaces were negligible. The English mathematician John Wallis went even further in defining (1656) an infinitely small quantity as $\dfrac{1}{\infty}$ (the first appearance of the symbol for infinity). Wallis illustrated his method by the calculation of the area of a triangle of altitude A and base B : if it is divided into strips $\dfrac{A}{\infty}$ wide running parallel to the base, its area is

$$\frac{A}{\infty}\,(1+2+3\ldots+B)$$

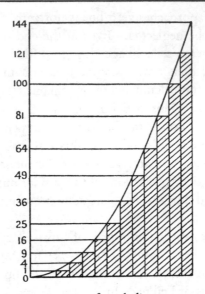

FIG. 6. Area of parabolic segment

Since the number of terms in this series is infinite, the expression reduces to

$$\frac{A}{\infty} \cdot \frac{\infty}{2} \cdot B = \frac{AB}{2}.$$

Though the method seems transparently absurd here, Wallis was able to apply it with success to far more complicated quadratures.

While avoiding such doubtful expedients, neither of the founders of the calculus offered a completely adequate definition of a fluxion or a differential. Leibniz took refuge in analogies, likening the infinitesimal to a grain of sand infinitely little in comparison to the Earth, and yet infinitely smaller in comparison with the whole universe. For long the greatest achievement of mathematics rested on a shaky foundation. That the calculus could be made to work at all depended, however, on the development of mathematics as a whole, for the way towards differentiation and integration had been well prepared in the years between 1640 and 1670. The problem of the curvature of curves had been investigated by numerous mathematicians of the mid-century period before Newton and Leibniz discovered a quite general treatment, so that their discoveries leaned heavily on the work of Fermat, Roberval, Barrow (1630–77), Slusius (1622–85), Mercator (fl. 1640–87), and Wallis, as they themselves were the first to acknowledge. The development of the calculus was a great co-operative enterprise, for whose success no one mathematician deserves the sole credit.

Although seventeenth-century mathematical history was so closely allied to that of mechanics, the calculus was barely applied to any part of science before the end of the century. Newton's first notes on his new methods date from 1665; in the following year he discovered the binomial theorem which rendered the calculus fully practicable. The earliest paper he is known to have circulated among his friends was *De Analysi*, written in 1669, which was followed by a new *Method of Series and Fluxions* (1671) and other pieces. Through these manuscripts, which Newton steadfastly refused to publish, and perhaps also through the lectures on mathematics which he gave at Cambridge in the years 1673–83, his discoveries became known to the most active English mathematicians, and one or two abroad. In 1687, in the *Principia*, Newton described for the first time in print the principle of fluxions, including the result

$$\text{if } y = ax^{\frac{m}{n}}, \ \dot{y} = \frac{m}{n}amx^{\frac{m-n}{n}}\dot{x}.$$

(Conversely, Newton was well aware that the integral of $ax^{\frac{m}{n}}\dot{x}$, that is, $\frac{n}{m+n}ax^{\frac{m+n}{n}}$, expressed the area under the curve $y = ax^{\frac{m}{n}}$; a type of quadrature familiar since the time of Fermat.) After this—partly owing to the publicity given to Newton's brilliance by Wallis in his *Opera Mathematica* (1693–9)—more interest in fluxions was manifested. Yet it was not until 1704 that the first of Newton's purely mathematical treatises (on quadratures) was actually printed.

By this time the calculus of Gottfried Wilhelm Leibniz (1646–1716) was fairly well known on the continent. Although his first and greatest interest was philosophy, Leibniz was drawn to mathematics through meeting Christiaan Huygens (1629–95) in Paris in 1672. He applied himself to the current problems of drawing tangents to curves and accomplishing their quadrature ; by 1675 he had begun to formulate the general methods of the calculus. It is unnecessary to enlarge on its differences from the Newtonian fluxions. The chief is that fluxions are velocities and differentials are magnitudes, so that $dx = dt.\dot{x}$.* Leibniz's notation and usage were easier to master and to handle, though his first account in the *Acta Eruditorum* of 1684 was not very enlightening. But Leibniz, with three years' priority in publication over Newton, was lucky. He explained and illustrated his methods in a succession of papers. Within a decade he had won the support of such very able continental mathematicians as the brothers Jakob (1654–1705) and Johann (1667–1748) Bernoulli, the Marquis de l'Hôpital (1661–1704), and Pierre Varignon (1654–1722). The Bernoullis revealed the power of Leibniz's calculus in solving mechanical problems, de l'Hôpital wrote the first systematic treatise explaining it (*Analyse des infiniments petits*, 1696), while Varignon reconstructed the whole science of motion in the new mathematical language. A generation of mathematicians devoted themselves with enthusiasm to developing the calculus and thereafter the mathematicians and astronomers of the eighteenth century confirmed its status as the basic language of physical science. By contrast, Newtonian fluxions, almost unknown still in 1700, were restricted to England until in the early nineteenth century " the language of pure de-ism " replaced the " dot-age " of fluxions. By this time, it was clear that the fluxional notation was less convenient, and that its concepts lacked definition ; moreover, all the great mathematical work for a hundred years past had been written in Leibniz's language.

* Obviously, if $dy = dt.\dot{y}$; $\dfrac{\dot{y}}{\dot{x}} = \dfrac{dy}{dx}$.

The really great revolution in seventeenth-century mathematics came too late to affect any of the major accomplishments of seventeenth-century science, even the *Principia*. By a strange irony, this synthesis of the new views of nature was presented in a mathematical form that belonged rather to the past than to the future. Its geometrical demonstrations were soon to be discarded by all serious mathematicians. Yet Newton himself had devised the fluxional calculus ! The paradox is to be explained partly by the circumstances of the time, partly by the surviving force of the geometrical tradition, but perhaps still more by the fact of Newton's own great facility in using those methods which he himself helped to render outmoded. There is a tradition—for which Newton himself provided authority[11]—that the propositions of the *Principia* were discovered in the first place by the use of fluxions, then demonstrated geometrically. In one instance indeed Newton arrived at a solution of a problem by fluxions which he was unable to prove in the geometrical form of the *Principia*. But this seems to have been unusual. In all the vast mass of Newton's papers there is, of course, ample evidence of his use of fluxions and other novel methods in pure mathematics ; of analysis by the same advanced methods of the problems treated in the *Principia* there is some trace, but rather little. Curiously enough there are attempts to solve problems by fluxions, apparently abandoned in failure, which Newton was easily able to solve by geometry. It seems likely that even in composing the most difficult mathematical-physical treatise of the age Newton relied only occasionally, perhaps in his most intricate investigations, upon a fluxional analysis. For the most part he could establish his results in the form in which he intended to demonstrate them.

What is true of Newton cannot be less true of other mathematical work in physics of the same period, since none other had, by 1687, an equal command of the new methods.[12] Certainly in the writings of Christopher Wren (1632–1723), Wallis and above all of the celebrated Dutch physicist Christiaan Huygens there was much mathematical practice that would have bewildered Galileo. Many of Newton's contemporaries used the algebraic notation derived from analytical geometry, and sometimes less formally geometrical proofs than those of the *Principia*. (However, Huygens' major published works on theoretical physics, *Horologium Oscillatorium* (1673) and *Traité de la Lumière* (1690), are fully as geometrical as Newton's book.) But as they lacked the calculus their writings— like the *Principia*—appear cumbersome and tedious to a modern mathematician.

Absence of elegance and fluency is clear evidence that mathematical resources are overstrained. The seventeenth century had amply proved that the book of nature—or at least of physics—was written in mathematical characters. Meanwhile the language of mathematics was itself undergoing a profound revolution. At the end of the century the differentials of Leibniz were ready to take the place of the triangles and circles of Galileo, conferring on the mathematical approach to natural phenomena a greater freedom and brevity than it had ever possessed before. In Newton's generation the problems of mathematical physics far exceeded the capabilities of such a geometer as Galileo had been—indeed, they required a Newton for their solution ; now physical science was about to exploit the potentialities of a new language of nature, which itself would make the problems facing Newton seem not too difficult.

CHAPTER IV

THE METHOD OF SCIENCE

To be possessed of a vigorous mind is not enough : the prime requisite is to apply it rightly.[1]

In an insular tradition that flourished for over two centuries Englishmen complacently regarded Francis Bacon (1561–1626) as the "Legislator of Science" (to use William Whewell's phrase). According to this tradition Bacon had provided the original impulse for a new effort in science by castigating the defects of existing knowledge, and by his discovery of the true principles of inquiry into nature had offered the method for remedying those defects. The history of modern science was regarded as the unfolding of Bacon's plans. The exaggerations of this opinion are patent and it has long been discarded. Bacon was opposed, in fact, to some of the chief currents in modern science, notably that stemming from Copernicus ; he did not instruct his contemporary pioneers in active scientific work, neither Gilbert nor Harvey in England, nor Galileo nor Kepler abroad ; his methods were not those by which the great achievements of seventeenth-century science were brought about. Nevertheless there are elements of truth in the Baconian myth. Its premise that the scientific revolution was an intellectual revolution was correct, and so was its realisation of the importance of new scientific methods. It was—up to a point—justified in emphasising the experimental character of these methods, and certainly in regarding Bacon as the most vocal exponent of experimentation. Above all, the Baconian myth was a seventeenth-century myth, not fabricated by historians but believed implicitly by all those Englishmen of Newton's time who took any interest in science. The founders of the Royal Society and many others before 1660 looked on themselves as disciples of Bacon. When they rationalised their idea of scientific method they invariably paraphrased him, so that even mathematical physicists like Newton could view their own work as an example of the famous inductive method. Nor was this high judgement of Bacon's teaching limited strictly to England. His continental reputation was less, it is true, but he was read abroad in the seventeenth century and quoted wherever sceptical organised empiricism was regarded as a virtue. The difference was that continental comments on Bacon were more likely than English ones to point out (as did Mersenne) that science could not be built upon empiricism alone. Yet Pierre Gassendi (1592–1655) and a few of the most experimentally minded continental scientists spoke of Bacon in almost as flattering terms as their English contemporaries.

Mersenne was right, of course. The scientific revolution was not effected by empirical methods only nor indeed by the use of any single new method of science. It was the replacement of one system of thought by another, and this process had begun long before the time of Bacon, or Descartes or even Galileo. The critical feature of seventeenth-century science was that it embraced new or revived ideas—an infinite universe in which the sun was but one star like countless others and life perhaps a commonplace, atomism, the mathematisation and mechanisation of natural processes, the idea of law and regularity—which were not and could not be proved experimentally and were related only incidentally to the rise of empiricism. If anything, empiricism was adopted because it offered some promise of verifying these ideas. To the metaphysics of the new science Bacon indeed contributed very little ; he was almost too late to help form them.

To the science of experiments, then, stood opposed—at least on a superficial inspection—the science of ideas. The Copernican answered the naive empiricism of Bacon by maintaining that the idea of the Earth's motion was more harmonious, more constructive than that of its stability, and so was to be accepted even though experiments could declare neither for nor against it. Similarly Galileo's attachment to the Platonic idea of the mathematical architecture of nature might be contrasted with the Baconian method of induction from experiments. Had Galileo been a Baconian he ought to have compiled from actual trials tables of the times taken by bodies to fall different distances and found the law $s = \frac{1}{2}at^2$ in his figures, just as Boyle's Law was found in a table of the compression of air. But Galileo had not done this. That Galileo had not proceeded in Baconian fashion did not however render his results suspect to intelligent Baconians, who observed that Galileo had empirically verified his theory in the end. Boyle's Law and Galileo's law of fall were equally acceptable truths in the last resort, however dissimilar the mode of their origin.

Implicitly at least the seventeenth century recognised two distinct methods of science, the mathematical-deductive method of Galileo and the inductive method of Bacon. It went still further counter to empiricism in allowing the use of reasoning from unverified hypotheses if they seemed plausible and in keeping with the general tenets of the new structure of scientific thought. For seventeenth-century science was Cartesian, as well as Baconian and Galilean. In fact the ideas widely held between 1660 and 1720 on how natural processes work owed far more to Descartes than to any other single person.

When the tendency of the time was to proclaim the merit of some single clue to the search for truth in science, to be found in reason or mathematics or experiment, it is not surprising that confusion of mind resulted. Far from having clear ideas about the route to the expanding certainty they desired philosophers and scientists were muddled in their language and in their accounts of scientific methods. Their accomplishments sprang not from a common epistemology but from the interplay of a multiplicity of notions. At times both Galileo and Descartes wrote like Baconians, while on the other hand Bacon's own best known exemplification of an inductive investigation—into the nature of heat—is nothing more than an instance of how one might compile a justification of a preliminary hypothesis.[2] (As an example of inductive proof it is worthless.) Such confusion about exactly what it was he was doing in order to arrive at a scientific proposition is evident in the writings of every seventeenth-century scientist. No label is universally valid for any of them. Cartesians did excellent experimental work ; those who considered themselves Baconian were quite as likely to rely heavily on Cartesian conjectures. Newton himself did not escape the confusion of thought. He declared magisterially *I do not feign hypotheses* ;[3] he wielded to tremendous effect both the experimental method (in his early researches on light and colour) and the mathematical-deductive method (in his theoretical mechanics) but—as everyone knows—Newton was also the author of innumerable scientific hypotheses. Newton's thought grew by consideration of hypotheses, just as Descartes' did, and it was a hypothesis that led him to discover the law of universal gravitation. But he did not·leave it there.

The apparent contradiction between Newton's declaration and the content of his own writings is, of course, a result of epistemological confusion, not scientific confusion. It would be false to suppose that Newton was muddled about his own conduct of a particular inquiry. He was well aware, for instance, that his account of the action of " fits " of easy transmission and easy reflection on the passage ·of light through bodies was a hypothesis (p. 269), not of the same standing as his discoveries about refrangibility. His difficulty was to know how to describe his own procedures and manner of reasoning in science—which were of more than one kind—and to distinguish them from the procedures of Descartes, rightly considered by Newton to be different from (and inferior to) his own. What Newton meant to condemn was not the formulation of working hypotheses to be confirmed or denied by the advance of knowledge, but the use of unsubstantiated hypotheses as the very groundwork of physics. It is typical of the state of seventeenth-century thought about science that he was unable to do this clearly. In order to show that science was not merely a fairly consistent body of ideas—a set of connected hypotheses that might or might not be valid—he had to speak like a Baconian (although Newton's physics was not Baconian) and to move still further towards a positivist attitude to knowledge. In so doing he contradicted his own attitude as a theoretical scientist, which was certainly not positivist.

Newton's contemporaries all faced the same problem. They knew what they were about, but how were they to express the complex truth that science consisted both of ideas about the structure of nature (which might be far from demonstrable) and of verified experimental results ? They had already begun to speak of scientific laws, of laws of nature. It was possible, indeed common, to suppose that the laws had been laid down at the creation by God. God, said Newton, could have created the world differently had he chosen to do so ; then the laws of nature would have been different. If this explained why constant and profound regularities in nature are discoverable, it did not explain how scientists do discover them. One might postulate that the laws of gravity would apply to any possible planet of Sirius, but certainly this was something that had never been tested. On the other hand laws of nature could hardly be reckoned as self-evidently true, while to regard them as axioms seemed to deprive them of their firm roots in experience. On the whole seventeenth-century science was content to approach such problems rather naively, leaving their detailed analysis to subsequent philosophers. As a result the statements of scientists varied according to the exigencies of argument. When the necessity for the proof of facts was in question they referred to Bacon. When they wished to reason and even speculate freely they followed the example of Descartes. And for the logical connections between facts and theories they could turn to the arguments of Galileo.

It is when one considers the strong influence of René Descartes

(1596–1650) on seventeenth-century science that the fallacy of the Royal Society's admirable motto *Nullius in verba* becomes conspicuous. Words indeed were still powerful. Many of the Society were themselves, in varying measure, adherents to the Cartesian philosophy of nature, and though Newton himself opposed Descartes directly on almost every issue, it would be absurd to deny Descartes' strong influence upon him. Newton's greatest work by its very title shows their relation : it is the *Mathematical Principles of Natural Philosophy* as against the *Principles of Philosophy* that Descartes had published in 1644—principles which Newton held to be thoroughly unreliable.

Though there were at least three major strands in seventeenth-century scientific method, during its middle period that traceable to Descartes was strongly ascendant. His influence overshadowed everything else, dimming the lustre of Galileo, obscuring almost wholly the discoveries of Kepler, and (outside England) under-cutting the empiricism of Bacon and the early English scientists, Gilbert and Harvey. Not before the second half of the eighteenth century did Cartesian scientific ideas altogether lose their command. It might seem from this, and from the ambition of Descartes to figure in his age as a second Aristotle, that his whole effect was retrograde. Baconian historians—while allowing exceptions for mathematics and optics—have written as though it was. No mistake could be greater. Though one recalls the long Newtonian struggle against Cartesian vortices, and Huygens' phrase that Descartes was the author of *un beau roman de physique*, it is a major historical error to admire Descartes only as author of the *Discourse on Method* (1637) and its scientific appendages, while dismissing his other scientific writings as the source of imaginative delusions.

On the contrary, they were the source of powerful inspiration, for in the first place Descartes offered a new principle of certainty to those who found the geometrical logic of Galileo too abstruse and narrow and the experimental method of Bacon too pettifogging, and indeed in its exemplifications trivial. To the grand questions then confronting science—does the Earth move ? is the universe infinite ? what are the natures of light and heat ? what is the cause of generation and decay ?—these other methods seemed to promise no immediate answer, hardly perhaps the road towards an answer. The old world as men's eyes had seen it had fallen apart and nothing had brought security in its place. The old philosophy was bankrupt; no one any longer had faith in the arguments and the tests of truth that had satisfied so many generations. Where was a fresh hold to which conviction could cling ? Descartes himself had followed the course of study that ended inevitably in doubt and uncertainty ; when, he wrote, he had completed such studies as were usual among learned men

> I found myself involved in so many doubts and errors, that I was convinced I had advanced no farther in all my attempts at learning, than the discovery at every turn of my own ignorance. And yet . . . not contented with the sciences actually taught us I had in addition read all the books that had fallen into my hands, treating of such branches as are esteemed the most curious and rare . . . I was thus led to the liberty of judging all other men by myself, and of concluding that there was no science in existence that was of such a nature as I had previously been given to believe.[4]

Descartes, then, had learnt to reckon " as well-nigh false all that was only probable " and at length he " abandoned the study of letters and resolved no longer to seek any other science than the knowledge of myself, or of the great book of the world ". It was a decision that many other seventeenth-century scientists were to emulate.

We do not need to swallow the whole autobiographical record of the *Discourse on Method*. Descartes was not a complete autodidact, though the *Discourse* makes no mention of the early influences to which he was subject, most lasting perhaps that of Isaac Beeckman in the Netherlands, when Descartes was twenty-two. Like all men he was less original than he thought. But the spirit of the narrative is authentic and it reflects the character of his mind. Descartes examined the metaphysical foundations of modern science more attentively than anyone else in his time, because he judged first that existing philosophy was full of errors and second that philosophy was the essential basis for all natural science. He did not take his metaphysical system ready made, nor allow it to emerge merely as an implication of a view of what the facts of nature are. He plunged in boldly. For Bacon the problem of knowledge had been : How can we obtain information ? For Descartes the problem was : What is real and what is illusory ? Again, the Cartesian model of science was not an encyclopaedia or a natural history but a logical pattern :

> The long chains of simple and easy reasonings by means of which geometers are accustomed to reach the conclusions of their most difficult demonstrations had led me to imagine that all things, to the knowledge of which man is competent, are mutually connected in the same way and that there is nothing so far removed from us as to be beyond our reach, or so hidden that we cannot discover it, provided only that we abstain from accepting the false for the true and always preserve in our thoughts the order necessary for the deduction of one truth from another.[5]

Observing his own four primary rules of reasoning and combining scepticism with order, Descartes within two or three months found that at least mathematics could be made to yield its secrets :

> not only did I reach solutions of questions I had formerly deemed exceedingly difficult, but even as regards questions whose solutions escaped me, I was enabled as I thought to determine the means whereby and the extent to which a solution was possible ; results attributable to the circumstance that I commenced with the simplest and most general truths and thus each truth discovered was a rule available in the discovery of subsequent ones.[6]

Why not ? Mathematics is a formal science and in principle one human brain could contemplate every logical nuance that mathematical reasoning can ever embrace. More important, however, mathematics illustrated the way in which logical relationship could interlock through the whole possible range of knowledge, holding it together in absolute certainty. Thus Descartes went beyond Galileo's notion of the mathematical proof of a proposition in physics, such as the law of fall ; he contemplated the possibility—indeed the necessity—that all natural science should be held together in a logical framework, though not a mathematical one. Galileo had limited mathematical science to " demonstrating some properties of accelerated motion (whatever the cause of this may be) ", but Descartes saw the logical structure of the science of motion extending from the cause of acceleration through all its properties. Therefore, in Descartes' eyes, Galileo's propositions (because he had consciously set causation aside) were inevitably incomplete and inaccurate, as he wrote Mersenne ; to him, Galileo had only applied reasoning to an isolated fragment of the story.

The problem still remained : how was the natural scientist to obtain for the fabric of the physical universe principles as certain as the axioms concerning lines and numbers with which mathematics begins ? Since geometry does not start with empirical measurements of lines and angles Descartes thought that science too should draw its first principles from reason. And in this there was no great difficulty, for some physical concepts he took to be as obviously true as the foundations of arithmetic. One could apprehend that two bodies cannot be in the same place at the same time as clearly as that two and two make four. The proof that a vacuum in nature is impossible was as transparent to Descartes as the proof that the number of primes is infinite. Or again, that matter is extension was as clear and indubitable a definition as Euclid's famous definition of a straight line. Such considerations, essentially drawing an analogy between the mathematical logic that already existed and the logic of physics that Descartes was striving to create, brought him to his crucial declaration :

> I concluded that I might take as a general rule the principle that *all things which we very clearly and obviously conceive are true* ; only observing, however, that there is some difficulty in rightly determining the objects which we distinctly conceive.[7]

Descartes explained how he felt the force of this conviction from the consciousness of his own existence : *Cogito ergo sum*. Of his existence he could have no doubt, therefore it was true that he existed, and in general there could be no greater guarantee of the truth of a proposition than the fact that, when assailed with all the rational powers at his command, it still shone clear and distinct and therefore true. Such of course was pre-eminently the quality of mathematical theorems ; if a theorem is deduced from the axioms without paralogism it must be true. Descartes was extending the test of logical consistency appropriate to the theorems of geometry, to the propositions of natural science.

Yet, after all, this was not quite enough. To identify truth whether in science or mathematics with logical consistency would leave the first principles uncertain. (Though in Descartes' day the definitions of mathematics were not considered to be arbitrary.) On these the mind might err, however impeccable the deductions that followed afterwards. God was needed as the ultimate guarantor of truth and perfection and this for Descartes was the best proof of his existence, as certain as a demonstration in geometry.

> How do we know that the thoughts which occur in dreams are false rather than those which we experience when awake, since the former are often not less vivid and distinct than the latter ? And though men of the highest ability study this question as long as they please, I do not believe that they will be able to give any reason which can be sufficient to remove this doubt, unless they presuppose the existence of God.[8]

That is, God is the ultimate touchstone distinguishing reality from mere appearance, truth from falsehood. When once this is perceived, then,

> it follows that out ideas or notions, which to the extent of their
> clearness and distinctness are real, and proceed from God,
> must to that extent be true.

In the last resort human confidence in human reason could only derive from the certainty that " all that we possess of real and true proceeds from a Perfect and Infinite Being ".

All this amounts to somewhat less of a justification of *apriorist* science than it seems at first sight. Descartes did not believe (any more than Aristotle) that one could solve all the problems of the universe in the study by thinking about them (though he was rightly signifying that thought on these problems is as necessary as any other activity towards their solution). When he wrote that a scientific concept that is clearly and distinctly conceived is true, he meant this in relation to the evidence that men have. He knew very well that no one could have a clear and distinct idea of the formation of the rainbow without some knowledge of the refraction of light by water. Though Descartes appeared to throw empiricism out of the front door it crept in at the back, when " . . . observing that there is some difficulty in rightly determining the objects which we distinctly conceive ". Secondly, Descartes insisted heavily upon rationality :

> whether awake or asleep, we ought never to allow ourselves
> to be persuaded of the truth of anything unless on the evidence
> of our reason.[9]

Again, his position was of the same kind as Galileo's though more broadly stated. Descartes saw the structure of science as essentially theoretical and that anything breaking the logical continuity of this structure must be false.

In fact there was little in the philosophy of the *Discourse on Method* to which Galileo or even Newton need have raised objection ; nor did any mathematical or empirical scientist seriously criticise the three treatises—*La Dioptrique, Les Météores, La Géométrie*—in which the method was displayed. The target of attack was the system of nature displayed in the *Principles of Philosophy* (1644).* The reasons for the vulnerability of this work were twofold. In the first place Descartes there extended the test of validity by clearness and distinctness from the phenomenological level to the prime characteristics of the architecture of the universe. In the Method Descartes had written of the strength of intuition :

> By intuition I understand, not the fluctuating testimony of the senses, nor the misleading judgment that proceeds from the blundering constructions of imagination, but the conception which an unclouded and attentive mind gives us so readily and distinctly that we are wholly freed from doubts about that which we understand. Or, what comes to the same thing, intuition is the undoubting conception of an unclouded and attentive mind and springs from the light of reason alone ; it is more certain than deduction itself in that it is simpler, though deduction cannot be by us erroneously conducted.[10]

Descartes' own concept of inertia, Galileo's concept of acceleration, Newton's concept of gravity were in the beginning at least intuitive. Descartes was quite right to stress the psychological significance of the intuition of discovery, even if he mistook its absolute status. But these are examples of intuition as a psychological process playing on the facts of experience, to which it brings sense and order. In the *Principles of Philosophy*, however, Descartes conferred certainty upon intuitions not directly dependent on experience. There is no rational escape from the proposition that a body will not deviate from uniform motion in a straight line, or from rest, unless some force is acting upon it ; but the proposition that even God could not empty a vessel of all matter is not in the same class. The same applies to Descartes' contention that there are just three types of matter : why not two or four ? The high claim for intuition became, indeed, absurd, when only the first of Descartes' seven rules

---

* The biological mechanism of *L'Homme*, posthumously printed in 1662, was also rejected by many, but for different reasons.

of collision was found to be correct. Unguarded, unrestricted intuition slipped insensibly into sheer imagination and yielded no less " blundering constructions ". And Descartes almost confessed as much when he professed to write not of the universe and men as they are, but as they must be.

Thus, in the second place, the weakness of the Cartesian system of nature, as distinct from the Cartesian philosophy of science, was its shadowy relevance to experience. Any critic of Descartes could echo Bacon : where are the experiments to prove this doctrine ? The Cartesian theory of light, arguing from its postulated particulate motion, demonstrated indeed the verifiable Snel Law of refraction ; but in so doing Descartes further assumed that light travels faster in water than in air. By different reasoning, requiring that the velocity in water be less, Fermat equally arrived at Snel's Law. Here was a clear case where reason was inadequate ; only measurement could decide which line of thought was just. (The experiment could not be made before the mid-nineteenth century when Descartes—and Newton—were proved wrong.) In such a case, as neither Descartes' theory nor Fermat's conflicted with anything known, they simply cancelled each other out. Cartesian physics contained a host of explanations of this kind which, if they conflicted with nothing save perhaps plausibility, were confirmed by nothing. The empiricist could claim that for all his reason Descartes in the *Principles* had left knowledge of nature as uncertain as he found it ; his proclamation :

> I have ever remained firm in my original resolution . . . to accept nothing as true that did not appear to me more clear and certain than the demonstrations of the geometers had formerly appeared

had proved in the details of physical mechanism to be fraudulent. A student of geometry who applies ruler and protractor to Euclid's 49th proposition is a fool ; but it was not foolish to ask for more than logical neatness as a demonstration of the passage of screwed magnetic particles from one pole of a magnet to the other as a cause of " attraction ". A particular particle cannot be merely reasoned into existence ; there must be some empirical basis for its characteristics.

Nevertheless, despite the defects of his physics and cosmology (inevitably far more obvious and troublesome in 1680 than they had been in 1640) Descartes' insistence on reason and clarity did offer a point of stability at a time when all was in doubt. Experience and experiment might speak with contrary voices, the senses might be bewildered by unassimilable information on the rich variety of nature, but Descartes confidently proclaimed the capacity of human intelligence to discover laws, principles, logical relationships. His was an influence exerted against the ineffable mysteriousness of the universe. He held that amidst the proliferation of detail the mind could attain universal truths and justifiably have faith in them. This was the inspiring lesson of Descartes which Newton (for example) did not disavow, though he disavowed Cartesian science completely.

Before Newton, however, Cartesian physics and cosmology as displayed in the *Principles of Philosophy* and *Le Monde* (posthumously published in 1664) carried away the imaginations of scientists looking for a new model, for all their arbitrary character. Descartes offered an insight into the structure of the universe that was modern, bold, rational and absolutely comprehensive. As he wrote at the end of the *Principles* :

thus I can demonstrate by a very easy reckoning that there is no phenomenon in nature whose explanation has been omitted from this treatise. For there is nothing to be included among these phenomena, but what we can perceive by means of our senses. And except for the motion, size, shape and situation of the parts of each body, which I have explained here as exactly as I could, we perceive nothing external to ourselves by means of our senses save light, colours, odours, tastes, sounds and the sense of touch. Of all these also I have just proved that we perceive nothing outside of our own thoughts, except the motions, sizes and shapes of certain bodies. So that I have proved there is nothing at all in all this visible world, in so far as it is merely visible or perceptible, beyond the things I have accounted for.[11]

It is only fair to view this success in its context and not in the light of the later evidence of its fallaciousness. While agreeing with ancient belief in the stability of the Earth (a purely formal act of casuistry, however) and the impossibility of the vacuum, Descartes had accomplished the tremendous task of reconstructing the whole interpretation of nature *de novo*, following for the most part an argument that is clear and superficially plausible, incorporating most of the more recent scientific observations, and avoiding altogether the perplexing forms and qualities, sympathies and antipathies, and other-worldly influences of learned tradition. To the mid-seventeenth century, searching for a system, Descartes revealed a new logic of nature with surpassing effect.

All static preconceptions have disappeared from the universe of Descartes. They are replaced by a grand conservation-principle : the quantity of motion in the universe remains ever the same. But if matter and motion remain constant as a whole in this indefinite universe (Descartes reserved the word infinite for God), the possibilities of local variation are endless. Descartes made only one distinction between the matter in the universe and the space it fills, that the former is movable. A material body, apart from its normal but inessential properties of hardness, colour, weight and so on is simply a shape in space that can be transported hither and thither without loss of identity. Matter fills all space, but only its grosser manifestations are directly perceived by men. It exists in three species : that which forms gross bodies opaque to light, that which consists of very small round particles forming the heavens and transmitting light, and that which forms the sun and stars and is the matter of light. In other words ordinary matter, aether and material light. The latter, lacking particulate form, is plastic enough to fill everywhere the gaps between the particles of matter and aether so that no space is left void. But the particles are not atoms, for they are infinitely divisible. Since the universe is completely full motion is a displacement or rearrangement, involving a constant impact of particle on particle which is the basis of Cartesian physics.

Under these conditions any movement tends to create a swirl or vortex. The solar system is in fact such an aetherial vortex with the sun at the centre and the planets swirled around it by the aether, some of them at the centre of subsidiary vortices carrying round satellites such as the moon. The whole universe consists of such vortices, each with a star at the centre, fitting together like a mass of soap-bubbles. Comets are celestial bodies that travel endlessly in the periphery of these vortices, passing from one to another. Since the aether of each vortex is rotating it tends to recede from the central star, exerting a pressure on contained bodies that is perceived as light radiating from the star and reflected from planets and satellites. The transmission of this pressure, or light, is of course instantaneous—in spite of what Descartes had postulated in connection with refraction. The spots on the sun are amalgamations of coarser particles floating like scum on its surface ; should these accumulate sufficiently they would form a skin of ordinary matter, the emission of light would cease, and the vortex collapse. Thus in time a star may become a planet and be captured as a passive body in some neighbouring vortex.

Descartes recognised as a consequence of his second " law of nature "—the first accurate statement of the law of inertia—that a planet would always tend to move in a straight line, escaping from its orbit in the vortex unless something held it back. So he explained that the aether of the vortex, being confined by its perimeter, would tend to force the planet inwards. Hence its position in a fixed orbit results from the opposition of the planet's own centrifugal force to the centrifugal force of the aether, reflected inwards, the size of the orbit being that which gives an exact balance.

There is much else on cosmology in the third book of the *Principles* that it would be tedious to review—why comets have tails,

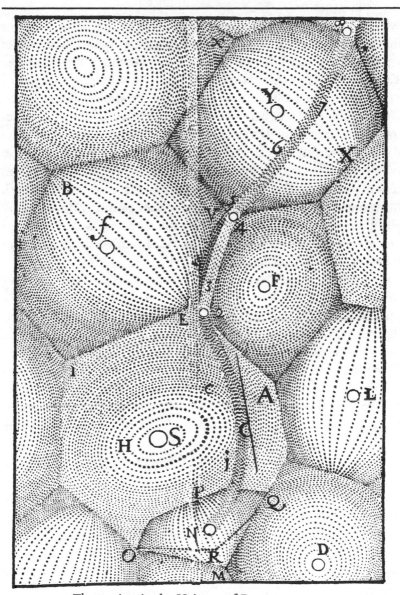

FIG. 7. The vortices in the Universe of Descartes. D, S, F, L, etc. are suns. The numbers mark the path of a comet. (From *Principia Philosophiae*, 1644)

why the moon always shows the same face to the Earth, why the inner planets move faster than the outer ones, why there is a kind of air round the sun. In the fourth book Descartes turned to the Earth itself, and physics proper. Gross earthly matter he regarded as totally inert, pushed about by various motions of the aether. Gravity, for instance, is caused by the tendency of aether-particles to revolve more swiftly than the sluggish Earth, so that they push down any earthly matter that can yield to them. The same action causes drops of liquid to assume a round shape, for the aether rushes round moulding them as the potter moulds clay on his wheel. Most ingenious perhaps is Descartes' explanation of magnetism by postulating two streams of screw-like particles, one with a left-hand thread and one with a right-hand thread, issuing respectively from the north and south poles and each circling the Earth to return to its pole by appropriately screwed channels through the Earth running parallel to its axis. In a highly complicated fashion these screw-particles act on iron and magnets (also supplied with screwed channels) to bring about the orientation of a suspended magnet and its attraction for iron.* Descartes left nothing untouched—the difference between transparent and opaque substances, the formation of minerals and the operations of chemistry, the force of exploding gunpowder, the tides, the formation of mountains, the hardening of steel by quenching, the attractive power of rubbed amber, the nature of glass and metals. . . .

The *Principles* was a triumph of fantastic imagination which happens, unfortunately, never once to have hit upon a correct explanation. Yet it was also a work full of creative ideas, if only because Descartes had observed faithfully the tenets of his own—and the seventeenth century's—mechanical philosophy. He was the effective progenitor of kinetic atomism, Daniel Bernoulli's mathematical treatment of gas-pressure (*Hydrodynamica*, 1738) being the

---

* Thus the problem of "action at a distance" in magnetism was overcome; "attraction" was the result of the mathematical action of particles. Others, including Newton, attributed it to a magnetic "effluvium".

link between Descartes and the nineteenth-century kinetic theory. The particular mechanisms described by Descartes were absurd, but the ambition to know them and the form he adopted for them were not. For this reason Cartesian science was taken seriously for a hundred years, and one can fairly describe such a scientist as Christiaan Huygens as a Cartesian, though he by no means subscribed to all the Cartesian explanations. Huygens looked at the world in the same manner as Descartes and expected to find the same kind of mechanisms in it, including the aether and the vortices. Indeed, it is impossible to open any book on physical science written between 1650 and 1720, not excepting Newton's *Principia*, without coming across the problems that Descartes had first tried to solve in the mechanical way, without recognising his shadow.

None of the basic principles of Cartesian physics was refuted in the seventeenth century, indeed none was decisively rejected before the end of the nineteenth. Newton was no less a mechanical philosopher than Descartes and, if he overthrew the Cartesian concepts of vortices and aether, nevertheless he fabricated alternative aetherial hypotheses (though he never committed himself firmly to them). For while Newtonian mathematical analysis showed that Descartes' hypotheses for explaining gravity, light, magnetism and so forth were quite fallacious, it did not—in the eyes of many philosophers and scientists—remove the need for such hypotheses. Aetherial and vortical notions were to have a long history in physics after Newton. Hence for all his fantasies Descartes must be reckoned among the great theorists of early modern science.

Perhaps because of this very fact the *Principles of Philosophy* sorely tried the patience of more empirically minded scientists even in the next generation. They too wished to find the causal structure of the world as simple as Descartes had imagined it to be. They too hoped to work back, by their painstaking methods of experiment, comparison and calculation to material particles and motions as the roots of things. But Descartes had cheated. He had not consulted their laborious notebooks nor had he really faced the intricacies of any single phenomenon of nature. He had failed to tie theory—though his was in the broadest sense the type of theory they believed to be useful—to the detailed facts of investigation; he had not even attempted to do so. Nor had he scrupled—this perhaps worst of all—to feign hypotheses which he did not believe to be true in order to fabricate an explanation. And his pretense of geometrical certainty ("I wish nothing to be taken for truth but that which is deduced with so much rigour that it could serve as a demonstration in mathematics"[12]) was by this fact alone made farcical. Descartes, in short, was so imaginative a theorist that he had jeopardized the principles of explanation and the very metaphysics which seemed valid to the empiricists themselves, by elaborations only too patently false.

The empiricists were perfectly right. Before we can discover why the world happens to be as it is we must first discover what it is. Descartes had known this well too, and had attended to the necessity more carefully than his book of *Principles* might suggest. In his letters he appears more often as the conscientious, inquiring scientist. But he regarded naive empiricism (the "thus I confute Berkeley" of Samuel Johnson) as fallacious, and the collection of incoherent "natural history" information as useless. The beginnings of knowledge are *a priori*:

> I remarked with regard to experiments that they become always more necessary as one advances in knowledge, for at first it is better to make use only of what is spontaneously presented to our senses, of which we cannot remain ignorant provided we bestow any reflection upon it, however slight, than to concern ourselves about more uncommon and recondite phenomena. For the more uncommon only mislead us so long as the causes of the more ordinary remain unknown, and the circumstances on which they depend are almost always so special and minute as to be highly difficult to detect.[13]

Experiment had its greatest value in Descartes' view as a means of decision between two or more theoretical possibilities, for he admitted that his principles did not yield specific, unique explanations of the phenomena of nature. Hence

> my greatest difficulty usually is to discover in which of these modes the effect is dependent upon them ; for out of this difficulty I cannot otherwise extricate myself than by again seeking experiments, which may be such that their result is not the same, if it is in one of these modes that we must explain it, as it would be if it were to be explained in the other.[14]

This is not very different from the attitude of those modern scientists who speak of the advance of theory making possible predictions of new effects, which in turn are verified or falsified by fresh experiments. The obvious deficiency of any such tightly logical view of scientific procedure is not that it excludes experiment, but that it fails to explain how anything quite new ever turns up, except as a result of reflection on what was known before. It leaves no room for that part of the task of science which is concerned with describing the world as it is, from which facts of great stimulus to the theoretical function have emerged.

Though we may give Bacon credit for stating the descriptive aspect of science in emphatic terms, once again he was merely summoning labourers to a field already under active cultivation. The critical evolution towards modern attitudes, if not towards modern methods, in such observational sciences as astronomy, anatomy and embryology, botany and zoology, geology and mineralogy had already occurred in the sixteenth century. The first step was to perceive that in all these branches of knowledge detailed and precise description of what is, was a worth-while and necessary activity of science. And this seems to be connected with two other ideas : the idea that there is an endless variety in nature (so that to know the morphology of one plant or the anatomy of one animal is by no means to know them all) ; and secondly, the idea that amidst this astonishing variety among animals and plants, rocks and earths, there is an order and arrangement that make possible classification and generalisation. Possibly descriptive science is never pure description. Certainly there was a tendency for problems of explanation to become far more central in the seventeenth-century treatment of the descriptive sciences. Anatomy, as always, was inseparable from physiology. The question of the formation of rock strata and of major features like sea-basins and mountain ranges, and above all the question of the origin of fossils, began now to press upon geological writers. Biologists were becoming seriously concerned with problems of the distribution of species, of spontaneous generation, and of variation ; they also brought the concept of a species as related descendants of an originally created distinct form under critical examination. It was less and less useful to compile natural histories without discussing ideas, principles and methods. It would be misleading to indicate that naturalists and anatomists were as yet developing a conscious philosophy in their studies, or that their main effort went towards explanation and correlation. But the tendency was apparent and increasing.

The burden of trying to work out what, and how much, the facts disclosed by either experiment or observation could be taken to prove in the realm of ideas fell more heavily upon the physical scientists than upon the biologists. But in any field of science it was possible to be too naively empirical. Robert Boyle (following an earlier precedent of van Helmont) grew pumpkins from seed in a measured quantity of dry earth, to which water alone was added. After some months his gardener could show some fine pumpkins while the weight of earth remained practically unchanged. Clearly (and the answer is only marginally but still decisively wrong) the pumpkins consisted of nothing but water. It was a false reading of the experiment to conclude (as van Helmont, Boyle, Newton and many others concluded from this and other evidence) that water was transmutable into earth—the dry matter actually recoverable from the pumpkin—even though there was no *a priori* reason why transmutation should not be effected. With the resource of hindsight it is not hard to see the fallacy of basing such a conclusion on an experiment in plant physiology, when the physiology of plants was precisely what was not understood. Such an instance shows that in order to obtain a clear, unambiguous answer from an experiment— such as Descartes sought—every one of its theoretical implications must be known. The alternative was to grope in the dark. How dubious, for this reason, were the chemical arguments that related to the destructive distillation of complex organic materials such as blood, horn, wood and urine must be obvious enough. It is fairly easy to decide practical questions by making a trial (will coal smelt iron as well as charcoal ?), but the conditions of an unambiguous answer to a point of theory make demands on the state of the theory itself. Empiricism alone does not make experimental science.

Most conspicuously is this true in seventeenth-century medicine, at once high conservative and high empirical. Ringing the changes on drugs, poultices, setons, bloodlettings, baths and dietary rules the treatment of patients was an unceasing experiment. Many physicians would claim that their preferred method in a particular disease was experimentally proved—patients were cured by it. One might expect that by selection the better remedies would drive out the worse, but nothing of the kind happened. Random experiments by thousands of doctors on millions of patients, conducted with an air of great professional assurance, produced no positive results. Did bloodletting cure fevers ? No one knew, though everyone supposed it did. In the absence of any relevant theory, or the use of statistics, experiment with this therapy or that was meaningless.

In situations more amenable to close analysis than those encountered in clinical medicine men also did not feel constrained to draw the same conclusions from given experiments. In his famous experiments on refraction described in a first paper in the *Philosophical Transactions* of 1672, Newton found that a spectrum of colours was formed by the spreading in one direction of the circular beam of light falling upon his prism, and that this spreading occurred because the rays showing violet in the spectrum were more strongly refracted than those showing red. Each colour possessed its own degree of refraction. "Colours", wrote Newton, "are not Qualifications of Light . . . but Original and connate properties, which in divers Rays are divers." Generally speaking the light emitted by luminous bodies was a confused mixture of rays "indued with all sorts of Colours", and if the mixture was appropriate appeared white. Newton did not consider that these statements embraced any speculation or hypothesis. As he wrote later to the publisher of the *Transactions* (and Secretary of the Royal Society), Henry Oldenburg :

> You know the proper Method for inquiring after the properties of things is to deduce them from Experiments. And I told you that the Theory wᶜʰ I propounded was evinced to me, not by inferring tis thus because not otherwise, that is not by deducing it onely from a confutation of contrary suppositions but by deriving it from Experiments concluding positively and directly . . . And therefore I could wish all objections were suspended taken from Hypotheses or any other Heads then these two ; Of showing the insufficiency of experiments to determin these Queries or prove any other parts of my Theory, by assigning the flaws & defects in my Conclusions drawn from them ; Or of producing other Experiments wᶜʰ directly contradict me, if any such may seem to occur. For if the Experiments, wᶜʰ I urge be defective it cannot be difficult to show the defects, but if valid, then by proving the Theory they must render all other Objections invalid.[15]

To Newton it seemed that his experiments would bear only one interpretation, that given in his new doctrine of light and colours. He was astonished that certain critics of his paper, without disputing the experimental results, nevertheless took issue with that doctrine as though the two could be separated. While accepting the facts, Robert Hooke had declared :

> as to his [Newton's] hypothesis of solving the phenomena of colours thereby, I cannot yet see any undeniable argument to convince me of the certainty thereof.[16]

And Christiaan Huygens thought that Newton " should content himself that what he has advanced should pass for a probable hypothesis ".[17]

Neither Hooke nor Huygens were inexperienced reasoners in science, indeed they were more experienced than Newton himself. Yet they could differ radically from him on a fundamental point of the interpretation of experiments. All three agreed that Newton had had certain visual sensations. All three agreed that these represented certain facts. But while Newton went on to regard the constitution of white light from a confused mixture of coloured rays as another fact indirectly made plain by the former ones, Hooke and Huygens thought that there might be an alternative explanation—that there was a better alternative explanation in fact—so that for them Newton had described only one possible theory or a hypothesis. The lesson from the experiments that seemed so easy and simple to Newton did not appear in at all the same light to them, for the very good reason that both Hooke and Huygens were already committed to a notion of the nature of light which they took to be incompatible with Newton's doctrine, though the latter attempted to show that this was not the case.

The lessons from experiments only appear clear and simple in textbooks or to those to whom the lessons are welcome and acceptable. Few if any experiments are utterly incapable of being interpreted in more than one way at the time when the making of them is significant. This is as true of contemporary as of seventeenth-century science ; but the seventeenth-century scientists had to struggle intensely with this problem of the interrelation of theory and experiment for the first time. They had moreover to compromise between their desire to formulate precise natural laws and sharp ideas, and the affirmation that such ideas and laws were experimentally valid. In January 1663, for instance, Robert Hooke made some measurements of the compression of a closed volume of air by a column of water that may be rendered as follows :

Reduction in Volume of Air (%)	Length of Compressing Column of Water (inches)			Theoretical Length (Boyle's Law)
	EXP. I	EXP. II	EXP. III	
0	0	0	0	0
4·16	20	17	13·4	1·71
8·33	37	36	32	36
12·5	58	58	53·6	56·5
16·66	79	80	78·4	79·2
20·83	–	105·25	104·1	104·2

Hooke noted that the " three observations being so different from one another, may seem to overthrow each other, and the certainty of this kind of experiment in general . . . it being almost impossible to make these trials so accurate, but that there will be some mistake committed ".[18] Quite rightly he thought some of the discrepancies in his figures were due to changes in atmospheric pressure, but even allowing for this it is clear that the various results do not agree precisely with themselves, still less with those calculated from Boyle's Law. Indeed, at that time no results could have been produced that would have confirmed that law better ; yet it was impossible not to believe that the errors were in the procedure, and not in the statement of the law. That the change in a volume of air should be directly proportional to the change of pressure upon it was so obviously sensible that no one could doubt it—and the impossibility of exact or universal verification was acceptable.

Nearly twenty years earlier Marin Mersenne had found the same difficulty in confirming Galileo's law of fall by repeating Galileo's experiments with the inclined plane. He too discovered that measurements exactly as predicted by the theory were not to be had, and was therefore led to doubt whether Galileo's own verification had been as strikingly successful as the *Discourses* claimed. Not that Mersenne was sceptical of Galileo's law ; the point to which he had drawn attention was that such an ideal law could never be upheld in the world of experience.

Practically, the obvious step was to prepare the means for

performing experiments with greater accuracy. Galileo's law can be more accurately demonstrated than by the means that he or Mersenne controlled, as was done in the eighteenth century. That step was taken; in the second half of the seventeenth century there was a marked increase in the care devoted to experimental research. The experiments now described were real ones that the writer had actually performed, usually several times. When they were particularly notable they were commonly duplicated by the members of the various scientific societies, either individually or in public session, and failures were fully discussed. In physical experiments measurement was more scrupulously carried out, as is evident from the work of both Huygens and Newton. A product of such attention to detail was accuracy in determining important constants of physics. The value of the gravitational constant was established by Huygens from pendulum experiments as early as 1664, and with high accuracy. From the work of Picard and other geodesists the size and even the shape of the Earth were deduced ; the velocity of light followed from the observations of Roemer on the occultations of Jupiter's satellites (1672–6)—though it was underestimated because the solar system was taken to be smaller than it is. For the first time experimental procedures were properly recorded, as in the minutes of the Accademia del Cimento and the Royal Society, and both papers and books were written in which the course of an experimental inquiry was carefully reported step by step. This form of presentation, favoured by the empirical tradition in England, was initiated by Boyle who for the first time took the trouble to describe laboratory science honestly and fully. In discussing some thousands of experiments in physics and chemistry he did not conceal his frequent disappointments, nor the difficulty of duplicating results, nor the doubtfulness of their interpretation. Similarly in Newton's optical writings the reader is never far removed from actual experiments and even the mathematical *Principia* contains much evidence of Newton's skill and persistence as an experimenter.

Most powerfully impressive, however, in experimental science after 1660, in the work of Huygens, Newton, Mariotte, Boyle, Hooke or Mayow, is the effective alliance at last of experiment and theoretical analysis. Indiscriminate, unplanned series of experiments were still frequent ; they were typical of the ordinary business of the societies in Florence, London and Paris where a few, often indecisive experiments on some theme would occupy three or four sessions until attention turned elsewhere. There were so many promising openings for experiment which could at least be expected to bring forward some unexpected facts. Nor is it surprising that such a constant experimenter as Boyle should have looked into many blind alleys. What the science of Newton's lifetime offered as profoundly original, however, as the strongest reaction to and most effective improvement upon the pure rationalism of Descartes, was the unfolding of theory through experiments, each dependent upon the other. In this way the exploration of nature became at once an empirical and an intellectual process, the hand and the brain not being divorced (as Bacon in the *New Atlantis* had imagined they might be) but linked together in intimate partnership. Hence experimental science became, by the manner of its evolution, at the same time descriptive and explanatory.

Bacon had taught that scientific knowledge could be built up merely by asking questions that experiment and observation were capable of answering. Though he had not solved the problem of discovering the right questions to ask, nor shown how the result could be more than an encyclopaedia of miscellaneous facts summarised by induction under doubtful generalisations, he had set experiment as the chief test to be employed in the descriptive function of science. *Fiat experimentum* was never to be an exclusive, automatic solution to any problem, but it proved an essential element in all solutions. Meanwhile, the contributions of Galileo and Descartes—both effectively posterior to Bacon—to the method of science were at a different and in a sense a higher level. The problem to which the mathematical philosophy of the former and the rationalist philosophy of the latter were addressed was the construction of scientific explanations. Reduced to bare essentials the empirical science of Bacon and the theoretical science of Galileo and Descartes did not conflict, for they were not concerned with the same things, nor with the same functions. If the inductive method alone was a poor tool for expounding the universe, equally deductive logic could only work on a sound picture of what the universe is like. Scientific concepts and laws of nature do not crystallise out spontaneously from natural histories, but neither can they be whisked up as pure abstractions. The defects of Galileo's physics and the ultimate debacle of Cartesian science illustrate the truth that scientific theories must satisfy not only the necessary conditions of logic but the contingent conditions of nature. The planetary orbits could be perfect circles, but they are not. The strength of the scientific method that united empiricism and theorisation—especially mathematical theorisation—in a joint attack on the complexity of things was that it dealt with necessity and contingency at the same time. It made the real world comprehensible through ideas. It need hardly be said that model investigations of this kind are rare. Historically science has developed by strange paths that know no rule, along which accident, intuition and prejudice have been guides on occasion. That is hardly the point. The point is that whatever the psychological and biographical aspects of scientific discovery, what is discovered can be organised into the pattern of reciprocation between theory and experiment that the late seventeenth century amalgamated, almost unconsciously, from varied and discordant traditions.

CHAPTER V

FLORENCE, LONDON, PARIS

The business and design of the Royal Society is :
To improve the knowledge of naturall things, and all useful Arts, Manufactures, Mechanick practises, Engynes and Inventions by Experiments—(not meddling with Divinity, Metaphysics, Moralls, Politicks, Grammar, Rhetorick, or Logick).
To attempt the recovering of such allowable arts and inventions as are lost.
To examine all systems, theories, principles, hypotheses, elements, histories, and experiments of things naturall, mathematicall, and mechanicall, invented, recorded, or practised, by any considerable author ancient or modern. In order to the compiling of a complete system of solid philosophy for explicating all phenomena produced by nature or art, and recording a rationall account of the causes of things.
All to advance the glory of God, the honour of the King, the Royall founder of the Society, the benefit of his Kingdom, and the generall good of mankind.[1]

That science became fashionable is one of the odder facets of seventeenth-century intellectual history. Probably indeed George Bernard Shaw was not far out when (in a very unlikely encounter) he made one of Charles II's mistresses treat Isaac Newton as a bumptious usher and another demand of him a love-philtre to restore the errant monarch's passion.[2] No doubt the ladies of Versailles had a similar low appreciation of scientific genius, and their smiling contempt was shared by most of the butterflies and puppies who hung around the courts of Europe, great and small. The strange thing is that despite the mockery of science that professional satirists eagerly encouraged there were men of rank and fortune like Robert Boyle and Christiaan Huygens who devoted their lives to it; and that members of the three respectable professions of church, law and medicine made it their serious avocation. Most surprising, perhaps, monarchs could regard the scientists' efforts with grave if uncomprehending benevolence.

Literature had escaped from academic bonds early in the Renaissance, theology followed its example at the Reformation, but science had hardly become a subject of interest to men of the world before the late sixteenth century. Early in the next, for a variety of reasons, the academic life no longer offered the sole environment in which scientific studies could be peaceably followed, as it had in the Middle Ages. On his estate, in leisure from his profession, perhaps best of all in the large cities which furnished opportunity for meeting with others of similar taste, a man could obtain the books he wanted and read and work as he wished, free from the petty restrictions still normal in the universities that impelled the great majority even of studious men to desert their walls. Newer ideas and freer exchange of views were to be found elsewhere ; university professors— commonly with all too much justice—were regarded as stuffy, dull and conservative. Just as the modern poet or novelist seeks his fortune in the literary circles of the great cities so did the seventeenth-century scientist, rendering the position of one who did not, like Newton (for many years), almost as anomalous as that of Mr E. M. Forster. Universities were expected to teach boys, not to be research institutes.

When Newton was born science (and almost every intellectual pursuit except classical scholarship) was thus in a sense homeless, without focus. The medieval centralisation of learning in Paris, Oxford, Bologna, Padua or Vienna had widely broken down, though less in medicine than in other studies. If men of science were observed to congregate in Rome, Florence, Paris or London it was not because of universities but because of the congenial company and manner of life they found there, and on account of the patronage bestowed by wealthy and discerning courts. If they needed support and encouragement they looked for it to the state or a wealthy patron, no longer to the university and the church.

Galileo, in the spirit of the new age, had deliberately and to the great offence of the Venetian Republic thrown up his academic post at the University of Padua in order to become Philosopher and Mathematician at the ducal court of Tuscany, in Florence. (He was not the first to succeed in the art of winning patronage, where Leonardo had failed : Kepler was already Imperial Mathematician at Prague, and Tycho Brahe had enjoyed more splendid independence at Hveen). A generation later the Tuscan court patronised a group of scientists, Vincenzo Viviani (1621–1703), Galileo's last pupil, among them, in the Accademia del Cimento. But that organisation only lasted for a decade, from 1657 to 1667. When the king of France was persuaded by his great minister Colbert to embark upon the same plan the Académie Royale des Sciences so founded lasted (through a complete reorganisation in 1699 and others in the French Revolutionary period) down to the present day. The first king of Prussia founded a similar institution in Berlin, being largely inspired by Leibniz who thought that the lack of encouragement to science was a fearful slight upon the German nation. And, because the energetic activity of a national academy had already become a minor aspect of national prestige, Russia followed ; the St Petersburg Academy attracted some of the best minds in eighteenth-century science.

It is therefore a trifle disingenuous to write of the early scientific societies as though they were offsprings of a natural and spontaneous desire among scientists for discussion and collaborative enterprise. This was a part of the reason for their foundation but not the sole one. The earliest group that can qualify as an academy, the Accademia dei Lincei (Academy of the Lynxes), was indeed a purely private organisation. It properly came into existence in 1609 and collapsed with the death of its founder-patron, Federigo Cesi, Prince of Aquasparte, in 1630. Galileo and some of his closest friends were members, but the interests of the Lincei were not exclusively scientific. For all their ambitious schemes there were no regular meetings nor any effective co-operative work, though a small inner circle was more closely associated with Cesi's special interest in natural history. Its Florentine successor, the Accademia del Cimento (Experimental Academy), was a plain product of court life. Leopold, brother of the Grand Duke Ferdinand II, retained from his lessons under Galileo a keen interest in science shared by the Duke. A superb collection of instruments was formed in the ducal palace— some of them Galileo's—which the academicians used in common and which is still partly preserved. Leopold participated in the meetings of the academy and its experiments; the members met at his summons and under his direction. Viviani was court mathematician, Francesco Redi and Nicholas Steno were court physicians, Alfonso Borelli and presumably others of the group were stipendiaries of the Duke. The Accademia del Cimento only existed for as long as it flattered the interest of the Medicis and enhanced their prestige; when Leopold became a Cardinal in 1667 the meetings ceased, as some said because they were not fitting to his new dignity or (as others alleged) because of the ceaseless quarrels among the academicians. It is only fair to add that Leopold continued to show interest in science and still tried to attract notable men to Florence.

The theme of many scientists in search of a patron appears again in France at about the same period. There are no definite records of early gatherings of scientific men in Paris, partly because some of the more eminent among them were rarely seen in the capital in the 1630s and 1640s. Descartes lived in Holland, Fermat at Toulouse, Peiresc at Aix-en-Provence (often accompanied by Pierre Gassendi), and others worked at the University of Montpellier. But if Paris did not yet dominate French science it was the clearing-house of information through the elaborate network of correspondence maintained by the Minim friar Marin Mersenne (1588–1648) from his convent near the Place Royale. Mersenne was not a great scientist. He had, however, a capacity for appreciating and reporting other man's work pretty accurately ; his criticisms were generally sound, and in his own writings (mainly on physics and applied mathematics) he reflected the trends of his time, moving from anti-Copernicanism to acceptance of Galileo's ideas and, with reservations, of the mechanistic philosophy of Descartes. His chief significance in history lies in the letters he exchanged with virtually every Frenchman who was at all active in science, with Galileo and others in Italy, Hevelius at Dantzig, Thomas Hobbes and Theodore Haak in England, and many more. Each correspondent enjoyed the benefit of Mersenne's shrewd insight into what was going forward in European science. Nor did Mersenne hesitate to provoke conflict between holders of opposed views in the presumed interest of truth. He thus furnished a strong if roundabout link between some thirty or forty scientists, mostly French, and his convent was to some extent a rendezvous for those who lived in Paris or came for a visit.

Other meetings, conferences or *bureaux d'adresse* existing in Paris at this time seem to have had only a tangential concern with science, their main business being the dissemination and discussion of political news. One such group, known as the Cabinet des frères Dupuy, which had a long and useful life from about 1615 until 1662 and was frequented by men of intellect and high social standing, was certainly able to follow the latest scientific and literary gossip as well. The principal scientific assembly in Paris did not take shape until 1654, six years after the death of Mersenne. It met at the house of Habert de Montmor (d. 1679), an ardent virtuoso and amateur of science who failed to become the patron of Descartes but succeeded in the same role with Gassendi. Probably Gassendi was the president over informal meetings there during the last year or so of his life. From 1657 onwards the meetings were regulated under a more formal constitution drafted by Samuel Sorbière (1615–70). Membership was limited to those " curious about natural things, medicine, mathematics, the liberal arts and mechanics ", for the object of the conferences was " clearer knowledge of the works of

God, and improvement of the conveniences of life ". Two members were appointed to read papers at each session ; comment and discussion took place in orderly fashion.[3] Soon the Montmor Academy was a fashionable place of resort. At a meeting in 1658 (when Huygens' letter announcing his success in elucidating the nature of Saturn's rings was read) there were forty persons present, among them " two *Cordon Bleus* . . . both Secretaries of State, several Abbés of the nobility, several Maîtres des Requêtes, Conseilleurs du Parlement, Officers of the Chambre des Comptes, Doctors of the Sorbonne ", besides the more humble amateurs of science and mathematics.[4] Science had become respectable and even interesting to some of the best Parisian society. Henry Oldenburg, soon to become Secretary of the new Royal Society of London for the Improvement of Natural Knowledge, was much impressed by the conduct of these meetings and the matters discussed, when he participated as a guest during his visit to Paris in 1659. On 28 June he wrote a long account of a meeting at which spontaneous generation was discussed, most of the speakers being incredulous ; and later he reported a list of varied topics under discussion such as

The Source of the Variety of Popular Opinions, The Explanation of the Opinions of Descartes, The Insufficiency of Movement and Figure to explain the Phenomena of Nature (undertaken to be proven by an Aristotelian). Then of the Brain, of Nutrition, of the Use of the Liver and Spleen, of Memory, of Fire, of the Influence of the Stars, If the Fixed Stars are Suns, If the Earth is animated, of the Generation of Gold, If all our Knowledge springs from the Senses, and several others, which I do not at this moment remember.

As one briefly in the company of Wilkins, Boyle, Wren and others at Oxford, however, Oldenburg was not wholly convinced of the virtue of the method of disputation ; " the French naturalists ", he wrote, " are more discursive than active or experimental. In the meantime the Italian proverb is true : Le parole sono femine, li fatti maschii."[5]

The group which met at the house of Habert de Montmor, and at those of others such as Melchisédec Thevenot (1620–92 ; famous traveller, and inventor of the spirit-level) and Jacques Rohault (1620–73 ; the chief systematiser of Cartesian physics) included other remarkable men besides these three : the mathematicians Pascal, Desargues and Roberval ; the astronomers Auzout, Bouillaud and Chapelain ; Clerselier, the editor of Descartes, and several physicians and chemists. They were frequently joined by visiting foreigners such as Christiaan Huygens, Jan Swammerdam the naturalist (also from the Netherlands), and the Danish anatomist Steno. But contemporary reports of scientific life in Paris (which was clearly brilliant around 1660) make it difficult to believe that the Montmor Academy was so closely organised as Sorbière's rules might suggest, and Montmor's personal connection with scientific matters lapsed in 1664. Among the Parisian scientists themselves there was dissension, partly over the merits of Descartes' work, partly over the importance to be attached to experiment and observation. Several of them wanted more than a debating-society ; they sought money and means for investigating scientific questions actively. Moved also by a desire to emulate the prestige and patronage which their colleagues in London had recently won— and no doubt greatly overestimating the financial measure of Charles II's benevolence—they began to think of gaining the approval of the greatest of all potential patrons, Louis XIV. No doubt they also reflected on the example of the Accademia del Cimento, to whom the court furnished splendid apparatus regardless of cost. The Parisians seem to have regarded the attempt to maintain experimental science in an academy upon private resources alone as quite hopeless. That, at any rate, was the purport of a memoir sent by Sorbière to Colbert in 1663, in which he asked for state help to preserve the Montmor group as a reformed, experimental society :

> to imagine that we might erect in this house a shop, a forge, and a laboratory, or to put it in a word, build an arsenal of

machines to perform all sorts of experiments, is not possible at all and is not the proper undertaking of a few private persons, although there are some very powerful in this company . . . Truly, only kings and wealthy sovereigns, or a few wise and prosperous republics, can undertake to set up a physical academy, where everything would pass in continual experiments . . . until the public is happy enough to meet Princes who have a taste for science, and for the perfection of the arts in use amongst us, or for the discovery of those which are lacking, our mechanics will remain imperfect as they are, our medicine will be blind, and our sciences will teach us only that there is an infinity of things that we do not know . . .[6]

Ultimately, having allowed the scientists to ventilate their own schemes for a couple of years, Colbert obtained Louis' consent to the foundation of the Académie Royale des Sciences. Though a less ambitious scheme than many had hoped for, it was the first instance of major state support for scientific research. Two rooms in the Royal Library were assigned to the use of the academicians. Appointments were made during 1666, sessions beginning at the end of the year. Few of the Montmorians and none of the outright Cartesians were chosen, though most of those appointed inclined to favour Cartesian science. Three of the most eminent early academicians, Cassini (Italy), Huygens (Netherlands) and Roemer (Denmark) were attracted from abroad to add to the lustre of French science. All sixteen were given salaries and somewhat later provided with excellent working facilities at the newly built Observatory.[7] On the whole they were left in peace to pursue science as they would, but little is known of the working of the Académie before 1699, when it was planned afresh on more positive lines.

The instinct to seek, and it must be added to confer, patronage was strong in the French scientific movement. It was prompted by the feeling that though scientists might do their own work well in isolation, the corporate investigation that was peculiarly necessary for scientific inquiry and the development of its applications needed larger sums of money than individuals could command. Because Louis XIV agreed, the organisation of science in France became if only loosely an aspect of the state. In England, where expectancy was just as great but unsatisfied, the Royal Society remained, despite royal patronage, essentially a private club. The French Academicians were appointed by the Crown ; the Fellows of the Royal Society elected their own colleagues. Yet the organisation of science on positive institutional lines had been more extensively canvassed in England than anywhere, ever since the publication of Bacon's *New Atlantis* (1627). England's fecundity in such schemes was occasioned by her political turmoils, in which a party triumphed that was not a slave to ancient academic traditions and that took a serious view of the responsibility of government for the moral, intellectual and material well-being of its subjects. The new men of the Commonwealth and Protectorate were presumed, not without reason, to be open to new ideas ranging over all topics from the union of the Protestant Churches to female education and the public reward of useful inventions. Scientific philosophers, who for the most part have ever been conformist in their political attitudes, were not dismayed by the Puritanism of their new masters nor even by their cutting off Charles I's head ; the thesis that the Puritan revolution actually stimulated science in England is a rather tenuous one (for scientists did even better under the sceptical Anglicanism of the restored monarchy), but it is certainly true that only a few scientists of intellectual distinction, Viscount Brouncker (1620?–84) and Isaac Barrow (1630–77) for instance, were driven out of the country by their own royalist principles.

In this atmosphere of present change and future hope religious enthusiasm merged into moral reform, and this in turn demanded a fresh approach to education ;[8] given a different and more Baconian twist the same code of ethics yielded the notion that it was a social good to alleviate the lot of the unfortunate who suffered under disease or the necessity for arduous physical labour. Given that toil and pain, hunger and death, are not virtues in themselves and that technological improvement is productive of social benefits rather than social evils, it was a small step to the Baconian argument that knowledge of nature confers power over nature ; and hence that to know scientific phenomena and laws is good. This philosophy did not first originate in England with the Puritan ascendancy nor was it held by all scientists, nor by Puritans alone ; and it did not cease to be influential when the monarchy and Anglicanism were restored. But it was a powerful philosophy during the decade of republican idealism (1645–55) although its most vocal exponent was a German Protestant, Samuel Hartlib (d. 1662). He was much less a scientist than Mersenne, of whom he is the insular shadow. His credulity was as great as his curiosity and his magnanimity. He was above all interested in machines, gadgets, recipes and formulas. He was a professional philanthropist who made it his business in life to disseminate information and inventions that he took to be useful though he had neither means nor capacity to assure himself that they were. In Hartlib the weakness of Bacon's social technology without the reinforcement of Bacon's philosophical alertness is very apparent, but he possessed an energy and charm attractive to more able men than himself. Some, like Boyle and Petty, were permanently influenced by him. While most of the Hartlib circle, dubbed by Boyle the " Invisible College ", are now quite forgotten, a few played prominent roles in the development of the Royal Society and English science.

No one has yet succeeded in disentangling completely the personal relations of all those who figure more or less largely in the scientific world of mid-seventeenth-century England. It is likely that most of the forty or fifty men concerned knew something of each other, though mainly associated with one of three chief groups —the Hartlib circle, devoted to social and ethical reform and more occupied with technology than abstract science ; or the club of mathematicians, astronomers and physicians meeting at Gresham

College ; or the Oxford Philosophical Society.* There were no barriers between them. Hartlib's associate Theodore Haak (1605–1690) was a leading member of the Gresham College club and was (according to Wallis) the first to suggest regular scientific meetings there. Boyle and Oldenburg moved from the Hartlib circle in London to John Wilkins' orbit in Oxford and the Philosophical Society. Many of the Oxford men also joined in the Gresham meetings from time to time.

Gresham College was the natural focus for scientific gatherings in London, since it had professors of geometry, astronomy and medicine regularly lecturing in the great town-house of its founder, the Elizabethan merchant-prince Sir Thomas Gresham (1519?–79). In the 1650s they were all men of ability. According to the recollection of the mathematician John Wallis (1616–1703), more than thirty years later, it was in 1645 that

We did by agreement, divers of us, meet weekly in London on a certain day and hour, under a certain penalty, and a weekly contribution for the charge of experiments, with certain rules amongst us to treat and discourse of such affairs ; of which number were Dr John Wilkins . . . Dr Jonathan Goddard, Dr George Ent, Dr Glisson, Dr Merrett, Mr Samuel Foster . . . Mr Theodore Haak . . . and many others.

These meetings we held sometimes at Dr Goddard's lodging in Wood Street (or some place near) on account of his keeping an operator in his house for grinding glasses for telescopes and microscopes ; sometimes in a convenient place in Cheapside, and sometimes at Gresham College, or some place near adjoyning. Our business was (precluding matters of theology and state-affairs), to discourse and consider of Philosophical Enquiries, and such as related therunto . . . We then discoursed of the circulation of the blood, the valves in the veins, the venae lactae, the lymphatick vessels, the

---

* These groups held aloof from political matters, though William Petty (an associate of Hartlib's) was a member of James Harrington's " rota club."

Copernican hypothesis, the nature of comets and new stars, the satellites of Jupiter, the oval shape (as it then appeared) of Saturn . . . the weight of air, the possibility or impossibility of vacuities, and nature's abhorrence thereof, the Torricellian experiment in quicksilver . . .[9]

All those named by Wallis except Foster, Gresham Professor of Astronomy who died in 1652, became Original Fellows of the Royal Society. Wilkins (1614–72) went to Oxford in 1648 as Warden of Wadham College, followed by Wallis in the next year when he became Savilian professor of geometry. Goddard (1617–75) also moved to Oxford where in the next few years they were joined by a brilliant crowd of younger men—William Petty (1623–87), Robert Boyle (1626–91), Christopher Wren (1632–1723), Robert Hooke (1635–1703) and Richard Lower (1631–91). These and others already at the University formed the Oxford Philosophical Society about 1650. Meanwhile meetings continued at Gresham College which by 1659 had again become the chief resort of those formerly meeting in Oxford. The Restoration brought in the Royalist scientists—without any apparent ill-feeling—Brouncker, Moray (d. 1673) and Barrow. A few months after the king's return, on 28 November 1660, occurred the memorable meeting at Gresham College at which (in the words of the Society's Journal Book) :

something was offered about a designe of founding a Colledge for the promoting of Physico-Mathematicall Experimentall Learning. And because they had these frequent occasions of meeting with one another, it was proposed that some course might be thought of, to improve this meeting to a more regular way of debating things, and according to the manner in other countryes, where there were voluntary associations of men in academies, for the advancement of various parts of learning, soe they might doe something answerable here for the promoting of experimentall philosophy.[10]

Within a week Sir Robert Moray had obtained royal approval of this new academy. The name Royal Society was first publicly employed by John Evelyn in November 1661 but it was not until 15 July 1662 that the Society was officially incorporated. Viscount Brouncker was the first President under the Charter, John Wilkins and Henry Oldenburg Secretaries, and Moray, Boyle, Petty, Wren, Goddard and other distinguished men were nominated by the Crown to the first Council.

The early schemes of Bacon, Evelyn, Petty and Cowley had envisaged an institution rather like an Oxford or Cambridge College—except that the Fellowships would not be sinecures. The members were to be resident stipendiaries, to teach as well as to investigate, and to be well provided with laboratories, workshops, observatories and chemical furnaces. Nothing like this materialised, though the Royal Society did not quickly lose sight of its ideal. Its resources, virtually, were no more than the members' shilling-a-week subscriptions—often in arrears. From Charles II it received honourable words and good-humoured raillery, but no money. It was barely able to support its Secretary, its Curator of Experiments (Robert Hooke), and an experimental " operator ". Indeed, had not the two first of these been singularly devoted men, the mainstays of the Society's existence, it could not have afforded to enjoy their services. The Society gradually acquired some apparatus, much of it made by Hooke and the operator, part of it gifts like Boyle's first air-pump and Newton's reflecting telescope. There was also a museum of rarities. But, though experiments were regularly demonstrated (Hooke seldom failing if often unpunctual) the Society was never able to function as a research institute. After a few years the ambition to do corporate research had faded, as the Society found its true function in transmitting, verifying and criticising the investigations reported by scientists at large.

In these first years, however, some deliberate effort was made to direct the growth of science. Lists of inquiries were drawn up from time to time for despatch to persons presumed able to answer them from their special knowledge. Experiments thought to be

instructive were committed to individuals or groups for perfor-
mance. Others were encouraged to compile " natural histories " of
various trades ; some were actually completed and published ; but
this Baconian concern for technology soon vanished. There was a
tendency (evident in Sprat's *History* of 1667) to presume that the
Society would endorse this or that experiment or theory as verified
or sound. As no strong organisational measures were adopted to
ensure that the Royal Society should act as an institution following
definite policies, rather than as a forum for the discussion of
scientific problems and investigations, and as it depended entirely on
the vigour of individual members for its continued existence, it
inevitably tended to become such a forum. Within ten years or so
the main business at meetings consisted of the discussion of topics
either raised impromptu by those present, or provoked by the
reading of a member's paper, or arising from a report of work
going forward elsewhere. A great deal of matter for consideration
was brought in by the energetic Secretary. Through his vast
correspondence abroad Oldenburg had an intimate familiarity with
what was going on in science in all parts of Europe ; through him
it became customary for many European scientists as well as the
English to report the results of their work to the Royal Society.
When (from 1665 onwards) their letters were published in the
monthly *Philosophical Transactions* after being read at the meetings
the modern form of scientific communication was born. Like
Mersenne, Oldenburg was not averse to increasing his budget of
news by asking one scientist to comment upon the work of another.
He has been much blamed for arousing ill-feeling by this procedure,
but it was common practice among the semi-professional publicists
of the time and was in a measure encouraged by the Society.

The Society's meetings were not wholly spent on talk, however,
and it was this that differentiated them from the proceedings of most
similar bodies. When a paper was read or an idea discussed the
matter was rarely dropped before some experiments had been
performed before the assembled company. If these were not

described in the original paper suggestions for them were sure to arise in discussion. Hooke, as Curator of Experiments, was normally expected to perform them, and he was one of the most prolific in suggestions for them too. If the author was present he would often offer his own experiments. In one way or another almost every experiment or observation that came before the Society was submitted to the test of repetition, if not at a meeting then by Hooke or others who would report their success with it. If the original report could not be duplicated the Society was not easily satisfied.

This was a system that worked well, providing a good balance between the need to check assertions experimentally and the desirability of free discussion. Broadly speaking the Society gave a fair estimation of the work brought before it and on many occasions it offered solid encouragement to men who might otherwise not have persisted in science. Sometimes it was too generous, for a good deal of time was wasted on trivialities and matters beyond the power of its investigation. The Fellows were hospitable to everything new and strange in natural knowledge, from strange autopsies (including one of a man who died of swallowing musket-balls to clear his guts) and Boyle's one-eyed colt to physiological experiments, microscopical observations, chemical experiments, mathematics and astronomy. The topics debated or subjected to demonstration changed in bewildering fashion from one Wednesday to the next. Views offered by the same or different speakers varied from shrewd comments or sensible suggestions for experiments to the more fantastic and superstitious explanations that still prevailed in certain dark areas of knowledge. Discussion at meetings often fell far below the high standard manifest in most of the Fellows' own publications—by no means everyone was up to the level of relevance and plausibility held by Boyle, or Hooke, or Lower or Newton— yet an energetic and purposeful spirit of inquiry was always apparent.

Reports of the proceedings during the earlier years of the Académie des Sciences have never been published, but it seems that they were much more formal. The Académie was a much smaller group than the Royal Society (which had soon enrolled 150 Fellows) having no amateur members at home and fewer correspondents abroad. For the most part business consisted of the reading of reports by the various members, or groups, on the work on which they were engaged ; which was sometimes chosen by themselves, sometimes appointed to them by the Académie. Examination of new inventions and certification of those it approved was a duty owed to the State, and the Académie also embarked upon a complete Natural History. Its astronomers formed a strong team, more fortunate than their London colleagues in possessing a fine observatory, built in 1667. They did much to improve the instruments and technique of their science. Enjoying state aid they were also able to promote some elaborate undertakings : Jean Picard (1620–83) went to the site of Uraniborg, Tycho Brahe's observatory, to determine its position more accurately in connection with the use of the measurements Tycho had made there; and with his colleagues he measured a meridian arc in northern France in order to compute the radius of the Earth. A younger astronomer, Richer, went to Cayenne so that from simultaneous observations on Mars made there and at Paris the distance of the Earth from the sun could be calculated. These two last enterprises yielded more correct figures than any available before.

The leading academicians had to submit plans for their labours to Colbert—or thought it politic to do so. Duclos, the chemist, simply proposed the analysis of everything into its elements (which he took to be salt, mercury and sulphur) ; on one occasion forty live toads were subjected to destructive distillation in pursuit of this programme. For many years, until the entry of Guillaume Homberg (1652–1715), in 1691, most of the chemical reports consisted of these tedious and useless analyses. Homberg, who was born on the island of Batavia and did not learn to read until he was twelve, had later become a pupil of Robert Boyle. Claude Perrault (1613–88) presented the plan for botany and anatomy, touching on several questions of plant and animal physiology : do plants, for instance, like animals, possess an organ that is the seat of the soul ? Two publications, *Mémoires pour servir à l'histoire naturelle des Animaux* (1671) and *Description anatomique de divers animaux disséqués dans l'Académie Royale des Sciences* (1682), signified the interest of the early academicians in natural history.[11] Huygens, speaking for the physicists, had an eye on practical issues. Besides experiment with the air-pump and on the force of impact he proposed to examine the expansive force of gunpowder and steam, and the application of wind-power. His main desideratum, however, which shows the influence of Bacon even in Cartesian Paris, was a comprehensive " history of nature ", chemical, physical and biological, to serve as a groundwork for a philosophy of nature " suivant le dessein de Verulamius ". None of these schemes was strictly adhered to and, apart from their concentration on astronomy, the researches of the members of the Académie were as varied as those of the Fellows of the Royal Society. Inevitably the study of optics and light, of chemical composition, of the vacuum, of blood transfusion, animal respiration and so on was pursued in parallel by the two Societies, with frequent cross-connections between them. What was done in Paris was soon reported in London, and vice versa.

The Académie had to pay a price for its greater resources. The Royal Society was utterly independent of the State, to which it was responsible only for the management of the Royal Observatory at Greenwich, founded in 1675. (The powers of the Society in this respect were so ill-defined that room was left for a serious quarrel between the Astronomer Royal, John Flamsteed (1646–1719), and the Society early in the eighteenth century.) As an institution it had no duties; on more than one occasion it dissociated itself from the special interest of the Crown in navigation.[12] After the first few years and despite voluminous propaganda (not usually originating with genuine scientists) about the usefulness of science to trades and public affairs it turned more emphatically towards pure knowledge. The Académie Royale des Sciences, on the other hand, was sometimes treated as the scientific department of State and had to cope with problems urged upon it by higher authority. Its mathematicians spent much time on applied hydraulics and the design of the water-works at Versailles, and on military matters. What was worse than such specific tasks—which were not numerous—was the perpetual consciousness that they were supposed to be engaged on some useful work.

The corporate efforts of the Accademia del Cimento, naturally enough, display a certain nostalgia for the great days of Florentine science under Medici patronage, in the days of Galileo and Torricelli (who had the ill luck to survive his far senior master by only six years). The dominant spirit was that of Vincenzo Viviani who did so much to collect Galilean material for preservation, and to vindicate Galileo's discoveries. He wrote the first life of Galileo, which is the source of the Tower of Pisa story and other anecdotes. Care was taken to verify experimentally a number of Galileo's assertions that the master had based on reason rather than actual trial. The thermometer, to whose invention Galileo had a claim, was much improved and there were many experiments made with Torricelli's barometer. A fair proportion of the work described in the *Saggi di Naturali Esperienze* (" Essays of Natural Experiments ") of the Accademia seems to have been carried out towards the end of its life, since (for instance) there are many repetitions of the experiments figuring in Boyle's book on *New Experiments Physico-mechanical touching the Spring of the Air and its Effects* published in 1660. Other instruments explored were the hygrometer and the pendulum—but the effective invention of the pendulum clock had been made by Huygens in Holland in the year the Accademia was founded (1657), and it was he who published its dynamical theory. Experiments were also made on heat and cold, magnetism, the incompressibility of water, and other popular topics of the age. The enterprises of the Academy were thus somewhat narrowly conceived and their influence was the less because the *Saggi* were only widely read when their contents were no longer novel or important (an English translation was printed in 1684, a Latin one in 1731, and a French one only in 1755). Moreover—at least as reported in the *Saggi*—the Florentine scientists were singularly timid in theorisation. Alternatively, this characteristic has been judged a strong merit, but it also meant that the significance of the scattered results they gained was (and is) hard to discern. To regard the Accademia del Cimento as the main fount of seventeenth-century experimental techniques in physical science is an absurd exaggeration, for exactly similar work that was better integrated with theory, and far better known to the world, was done in Paris and London and even in Germany. On the contrary, as the Accademia del Cimento was the shortest-lived of the major scientific societies so it was also the least influential. The writings of individual members such as Giovanni Alphonso Borelli (1608–79), Francesco Redi (1626–98) and Nicholas Steno (1638–86) were indeed of wide significance ; but these were their independent efforts, no less than those of Huygens or Roemer, Boyle or Newton.

There were also scientific societies in Germany but with the possible exception of the *Collegium Naturae Curiosorum* (" College of those Inquisitive into Nature "), which published a journal, they were, as Leibniz recognised, of no more importance than other local societies that flourished elsewhere. The *Collegium* in any case was interested in medicine and medical science only; indeed, a medical bias in scientific activity is discernible over the whole of northern Europe at this time. Few men of other professions there followed scientific affairs. The movement towards a national, subsidized scientific academy in Germany, headed by Leibniz towards the end of the seventeenth century, reveals plainly the *étatisme* of publicised science. Leibniz—whose own contributions to mathematics and philosophy can hardly be described as less than abstruse—could nevertheless argue that laboratories, botanic gardens, observatories and museums were essential to the development of science, and that only by providing these and remuneration for researchers could the ancient leadership of Germany in the crafts be restored. He spoke with contempt of those who " consider the sciences not as something essential to the welfare of man, but as a game or pastime ". But in this special pleading it was not the interest of man that was to be served (as it had genuinely been with Bacon) so much as the interest of the State, " tout ce qui entre dans l'œconomique ou mécanique de l'état civil ou militaire ". Not surprisingly Leibniz upheld strict mercantilist policies, the highest expression of State supremacy in the economic sphere.[13] Of course, if State support was the object desired this was a persuasive plea to make. Yet Sorbière had put it more generously in his memoir to Colbert.

Did statesmen really swallow the contention that science was useful ? In England, not at all ; Charles II and his ministers, sceptical of the technological resourcefulness of the Royal Society, made no demands upon it and gave it no money. Some research in military technology was paid for by the State in England, but it was not done by the Society. In France Colbert, who believed that all problems could be solved by administrative action, seems to have been deceived and probably disappointed in his hopes for utilitarian science. After his death (1683) Louvois descended on the Académie with vigour ; scientific inquiries, he announced in 1686, were made either to satisfy curiosity or to serve useful ends. The Académie was too illustrious and had too much serious business before it merely to indulge its inquisitiveness. " I understand by a useful research ", he said, " one that relates to the service of the King and the State ", and as examples suggested that the Académie discover the effects of mercury, antimony, quinine and laudanum in various preparations and analyse the new drinks tea, coffee and chocolate. The chief results of this intervention were that Bourdelin—a coffee-addict according to Fontenelle—destructively distilled three pounds of excellent coffee-beans (which he might better have drunk), while Perrault worked out a scheme for a two-stage cannon, an inner barrel containing powder and ball being propelled by a second charge from an outer one. After this conspicuous failure the academicians were allowed to go on much as before, for it seems that not even Louvois was cynical enough to strip them of their pretence of service to the State. Courts spent large sums on architects, painters, musicians and dancers so the maintenance of a few useless, but perhaps glorious, scientists was no intolerable extravagance. Patronage on the grand scale was an attribute of royalty—a fact not lost on the founders of later scientific academies.

That some scientists deceived themselves with Baconian claims is highly probable. Huygens, like Galileo before him, was confident that he had found the way to measure longitude at sea. Admittedly his clocks never behaved as well as he expected them to do; but he was careful to abandon this line of research (close to the interests of the maritime Dutch) while he was in the service of the French King. Agriculturalists expected their inquiries about tillage and crops, and the novel practices they urged on conservative farmers, to bring about a marked increase in the productivity of the soil. More rationally the medical scientists hoped that, in the long run, their work on anatomy and physiology and their efforts to be more scientific in the diagnosis and treatment of disease, would enhance the powers of their profession. Chemists too always had an eye open for useful new drugs—though they missed the anaesthetic effect of ether. Such ambitions, if premature, were laudable. They had no cramping grip on the activities of the scientific societies, nor on these of individuals; nor did they induce the subjection of science to State purposes. If discoveries for the good of mankind should come so much the better; meanwhile at any rate knowledge was steadily increasing even though little of it was immediately applicable. The Royal Society was particularly insistent that any useful discovery made known to it must be open to free use; it would have nothing to do with the endorsement of patent medicines or anything of that kind. However Louvois might huff and puff, curiosity was a more useful and creative incentive than utility in the privacy of scientific discussion within the academies and societies. All scientists (following the example of Bacon himself) preferred luciferous experiments to lucriferous ones;

if, again like Bacon, they sometimes seemed to speak with two voices, now claiming knowledge, now the welfare of mankind as their objective, it was after all because these two desiderata were ultimately identical. Knowledge—for science at least—was the only available road to the betterment of the human lot.

In reality seventeenth-century scientists like most of their successors were more preoccupied with the road than with its termination, and their attitude was not much altered by public pressures either from the State or from opinion. Men like Galileo and Huygens were more avid for the intellectual distinction of penetrating the problem of the longitude than concerned for profit or the consequences for practical seamen and merchants. Academies and societies did little to render scientific research more organised than it had been before and nothing to shift its direction. Galileo was proud of being a Lincei but his membership of that distinguished society is utterly irrelevant to a consideration of his intellectual development, and its effect on the whole history of science. Very little credit belongs to the societies for the work which such members as Huygens, Ray, Malpighi, Boyle, Newton and many more carried out ; if the Royal Society encouraged Leeuwenhoek to be a microscopist it did not confer his genius upon him. The Societies flourished upon the individual creative energies of their most gifted members, not vice versa. Their true function was not discovery. Rather it consisted of providing a means of testing both experiments and ideas, giving prominence to what was good and (less successfully) suppressing what was bad. They offered a forum where one man could measure his wits and his attainments against those of another, where he could learn what the problems of contemporary science were and with what methods and success they were being tackled. Thus the societies helped—but they were not alone in this—to raise and make more consistent the idea of what scientific research is about and how it should be conducted. Secondly, of course, the societies performed a valuable function in transmitting information, though their service in this respect was uneven and was to some extent supplanted by the journals that began to be published in the latter part of the century. None of these was officially the organ of a society though the *Philosophical Transactions of the Royal Society*, at first Oldenburg's private venture, gradually acquired that character as it was continued by successive Secretaries.[14]   The *Journal des Sçavans* (1665) and the *Acta Eruditorum* (1684) were wholly private and independent. Finally, the societies did much to elevate the place of science in the world, to interest and stimulate the multitude of ordinary men, and thus to create a scientific public.   They made science a profession by their prestige, assuming in the eyes of the world the same organisational importance that the Colleges or Faculties of law and medicine enjoyed.   But unlike these they were international in membership, and catholic in their interests, and these perhaps were the most unusual and most notable facts of their existence.

CHAPTER VI

EXPLORING THE LARGE AND
THE SMALL

Since we have no animals on Earth more than 18 feet in diameter, I beg Mr Hooke to take the trouble to see if he can distinguish an animal of one foot diameter at a distance of five miles, and let him judge from that whether he can hope to see living animals even in the moon, unless indeed they are incomparably greater in size than ours.[1]

Man, it has been pointed out, occupies an intermediate position in the scale of things. The size of his body and the dimensions most familiar to him lie about half-way between the scale on which physicists measure the fundamental particles of matter and the vastly greater scale on which astronomers measure the composition of the universe. Throughout most of his history he has been familiar only with that small fraction of the scale of nature that was immediately accessible to his experience, and so he tended to measure the universe by his own stature. Medieval philosophers might avow that the Earth is but a point in proportion to the magnitude of the cosmos but even so their notion of stellar distances was fantastically contracted. Copernicus, adopting estimates that Hipparchos had made nearly two millennia before his own time, reckoned the Earth's distance from the sun as about four million miles. Now, in order that the stars should be remote enough to reveal no parallactic displacement due to the Earth's annual motion that would be detectable by existing instruments, it was sufficient to suppose them merely a few tens of thousands of millions of miles remote from the Earth. It was this (nowadays) trivial distance that was regarded by Copernicus' contemporaries as a nonsensical, impossible inflation of the size of the universe. Previously it had been confidently supposed that the bounds of the physical heaven could not be much beyond the orbit of Saturn—a trifling fifty million miles from the central Earth. Kepler went beyond Copernicus. He tripled the sun's distance, and thought that the stars must be rather more than 240,000 million miles from the Earth.[2]   About mid-century the Bolognese astronomer Giambaptista Riccioli further reduced the measure of the solar parallax,* making the sun's distance greater than 24 million miles ; Flamsteed in 1672 found the parallax not more than 10 seconds of arc and the distance accordingly at least 84 million miles while at the Observatory of Paris J. D. Cassini in the same year found it to be 87 million.   Newton in his *System of the World* adopted Flamsteed's figure though in the *Principia* (Book III) he worked with one slightly less, and in general a figure of 80–90 million miles for the sun's distance was conventional from about 1680 till quite late in the eighteenth century.[3]

If the solar system thus grew in men's eyes the whole universe expanded even more swiftly, for while the base-line of measurement —the Earth's orbit—lengthened, instruments were improved and yet still failed to detect the much-sought stellar parallax.   About 1700 it was quite certain that the stars must be more than four times as remote as even Kepler had guessed, beyond a million million miles. Newton was willing to believe that first-magnitude stars might be 10,000 times as remote as Saturn, that is, at a distance of 80 million million miles (about 14 light-years).   From this distance, he calculated, the sun would appear rather less bright than Saturn does to us. This was a crude estimation of the true order of distance for the nearer stars and such notions of magnitude were not exceeded for two hundred years.   Yet a writer in the *Philosophical Transactions* for 1694 could record with awe his computation that light from the stars might take as long to reach the Earth as a West Indies voyage —a mere six weeks.[4]

It was easier to measure the vastness of the world than to observe

* The angle at the Sun subtended by the Earth's radius.

the smallness of the particles of which it is formed, though imagination could give them any conceivable dimensions. Copernicus and many others recognised that near-infinite smallness is a logical concomitant of near-infinite vastness :

As with those tiny and indivisible bodies called atoms which though they are not perceptible by themselves and do not when taken two or several together immediately form a visible body, yet may be multiplied until they join to form a great mass in the end ; just so it is with the place of the Earth. Its distance from the centre (the sun) is not comparable with the immense dimension of the sphere of the fixed stars.[5]

Philosophers could not yet measure the size of atoms, however, if such entities did exist, though they could hope that the improvement of microscopes would in the near future allow at least the grosser corpuscles in the structure of bodies to be seen.

If [wrote Newton] those Instruments are or can be so far improved as with sufficient distinctness to represent objects five or six hundred times bigger than at a foot distance they appear to our naked eyes, I should hope that we might be able to discover some of the greatest of those Corpuscles. And by one that would magnify three or four thousand times perhaps they might all be discovered, but those which produce blackness.[6]

Newton thought by this high magnification to resolve particles having a diameter of about five millionths of an inch. His own limit of computed magnitude (in the theory of the colours of thin plates) was one-tenth of this, and represents the smallest dimension that entered into seventeenth-century theoretical physics. The finest microscopes of Newton's age—the marvellous simple lenses of Antoni van Leeuwenhoek—had an extreme magnification of perhaps four hundred times, but their useful limit of visibility was much less. The smallest objects reported by Leeuwenhoek measured about 2μ across, or one ten-thousandth of an inch, roughly. It was a superb achievement, though the resolution at this power did not permit detail to be observed ; yet the particles that Newton hoped might be seen were ten to twenty times smaller still.

Thus the scale of things definitely measured by man multiplied about 500 times in the course of the seventeenth century, or if we take instead Newton's theoretical extremes of dimension, no less than a million million times. It was a fantastic, incredible enlargement of human horizons. But science could as yet grip only tentatively at the twin limits of size. Astronomy was still restricted, effectively, to the solar system ; the stars could be catalogued but not studied. The sole new fact discovered during the seventeenth century concerning them was that they could not be resolved into discs by the most powerful telescope ; in 1718, however, Halley discovered that stars have a proper motion. Even the one star that was relatively close at hand, the sun, did not yield much new information. The formation and appearance of its spots were carefully charted ; its true size was recognised ; and Newton was able to show that the gravitational attraction of the sun is about 230,000 times greater than that of the Earth. (Newton's theory also proved, of course, that the sun was a material body, not a ball of the mysterious " matter of light ".) The source of the solar light and heat remained an enigma provoking many hopeless conjectures. Newton may have been the first to suggest that the stars have a roughly uniform distribution in space : those nearest the sun being roughly equidistant from it and each other, the next more remote group at the same distance from the stars of the first group and each other, and so on. The first group appear as stars of the first magnitude, the second as stars of the second magnitude, and so to higher orders. On this theory there should be about 12 first-magnitude stars neighbouring the sun, 50 of the second magnitude, 110 in the third group, and so forth, and these numbers agree fairly well with observation. Others besides Newton assumed (as the simplest hypothesis) that the more faintly a star shines the more distant it is likely to be ; but this is not always true. In the late seventeenth century, too, it was generally agreed that the universe is infinite, space filled with stars stretching indefinitely far beyond the limits of visibility of even the most powerful telescopes, for Kepler was the last astronomer to conceive of the stars as massed together in a thin shell concentric with the sun. And so the Christian heaven lost its physical place in the universe for ever. Many believed that some at least of the stars were other suns surrounded by their own planetary systems, and that their planets also might well be inhabited. The plurality of inhabited worlds was an entertaining conceit and one that the Church was powerless to render dangerous ; those who talked of it had the air of appearing bold without imperilling themselves. For his own age it was classically and charmingly expressed in the *Entretiens sur la Pluralité des Mondes* of Fontenelle (1686), an immensely popular little book that passed through 31 editions in its author's hundred years of life.[7]

It was easy, as Auzout reminded Hooke, for imagination to outrun observation. Imagination was a faculty seventeenth-century scientists did not lack. But even if telescopes could be made to magnify so highly that an elephant could be seen in the moon—say at least 10,000 times—it was impossible to believe that the thickness of the Earth's atmosphere and other hindrances to perfect vision would allow them to be used. And in fact, of course, practical telescopes nowhere near approached such high powers. To obtain glass for lenses that was not flawed was almost impossible and none could be made more than a few inches in diameter. The light-gathering power of such telescopes was low. To increase magnification and to reduce the aberrations of uncorrected lens-systems telescopes of great length were made : 14 feet was usual for a small astronomical telescope by 1660 and later instruments increased from 50 to 100 and even 200 feet. The lenses were mounted in a square wooden tube or upon a long wooden beam, suspended by ropes from a mast. In the so-called aerial telescopes—those of greatest length—the rigid connection between objective and eyepiece was dispensed with altogether. Some of the best lenses were made by amateur opticians like Huygens and Hooke; among tradesmen the

Italian Campani (c. 1620–95) was most praised for the quality of his product. It proved impossible to give lenses the aspherical curvatures required by optical theory, moreover about 1665 Newton discovered the cause of chromatic aberration which could not be removed by any way of figuring the lens. Despairing of improvement in the refracting telescope Newton constructed with his own hands the first working reflector—earlier suggestions for such an instrument had been made by James Gregory (1638–75) and E. Cassegrain. The mirror, however, made at this time of a silvery copper-tin alloy, proved as difficult to work accurately as lenses and the reflecting telescope did not become effective in astronomy until about the middle of the eighteenth century.

The qualitative results obtained with the larger refracting telescopes employed after the death of Galileo were certainly striking. In 1655, for instance, Huygens solved the mystery of Saturn's rings. Earlier observers had been puzzled by the strange " wings " on either side of the disc of the planet which their instruments could not make plain, and which seemed to come and go. Huygens saw that Saturn was surrounded by a thin flat ring whose tilt towards the observer varied periodically. He also discovered the first satellite of Saturn. Four more were found by J. D. Cassini (1625–1712) of the Paris Observatory, who was also the first to see the division of the ring and to note the white polar caps on Mars. Before he came to Paris Cassini had studied the axial rotations of Jupiter and Mars, and the periods of the satellites of the former planet.[8] His observations proved that they obeyed Kepler's Third Law—important evidence later for the truth of the theory of universal gravitation—but the satellites never proved useful as a celestial clock for measuring longitude, in the way Galileo had hoped. Continuing this study of Jupiter's satellites Ole Roemer (1644–1710), also at Paris, remarked that the period of the innermost satellite was less when the distance between the Earth and Jupiter was diminishing, greater when that distance was increasing. Thus, as the Earth approached Jupiter successive eclipses of the

Astronomers observing a solar eclipse. The image of the sun is projected upon a screen, and the limit of the obscured portion carefully recorded. From Hevelius's *Machina Coelestis*

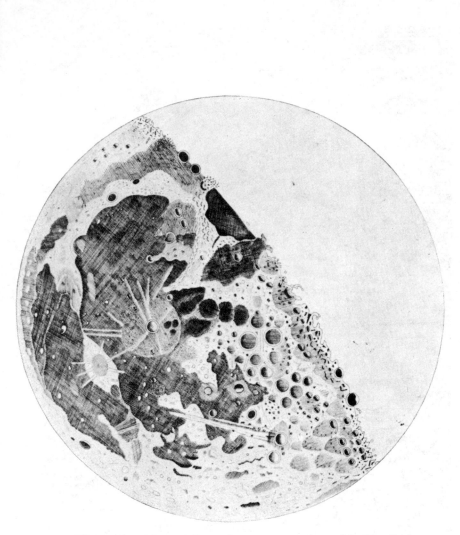

The waning Moon, 2 December, 1643, as engraved in Hevelius's *Selenographia*, 1647. This engraving conveys a good notion of the detail visible through telescopes of that period. Hevelius, using a Galilean refractor, plotted the topography of the Moon for the first time satellite occurred more rapidly than the mean, but when receding from the planet they happened less rapidly. Roemer accounted for this in 1676 by supposing that light travelled at a finite velocity. As the Earth moved away from Jupiter between two successive eclipses of the satellite the light had to travel farther, and so the interval between them was greater ; conversely as the Earth approached Jupiter. From his data he calculated that it would take light about 22 minutes to cross the whole orbit of the Earth. At first Roemer's explanation was treated with scepticism. Cassini attacked it. The old prejudice, first among the Aristotelians and then among the Cartesians, in favour of the instantaneous propagation of light was too strong to be overthrown at once, though Galileo and others had suggested that the speed of light might be finite. Huygens, however, and in England Flamsteed, Halley and Newton were convinced that Roemer's explanation was correct, though the latter reduced the time required by the sun's light to reach the Earth to seven or eight minutes.[9] Many others remained obstinate until James Bradley (1692–1762) discovered the aberration of light in 1729.

Roemer spent about six years in Paris, returning to his native Copenhagen in 1681. There he continued his observations, developing an important new instrument in the transit-telescope. This was carefully set to turn about a horizontal axis in the plane of the meridian, the transit of any celestial body across the meridian being watched by its passage across a web of wires placed at the focus of the eyepiece. This development owed much to Picard's earlier improvement of astronomical instruments in Paris. Picard (in 1668) was the first to attach a tube containing telescope lenses and cross-wires to a divided arc of circle for taking angular measurements. With his colleague Adrien Auzout (1630–91) he also devised a form of micrometer, consisting of threads controlled by a screw, which was inserted in the focal plane of a telescope eyepiece to measure the diameter of a planet, or the distance between two objects, in the field of view. Such methods as these, rapidly adopted by most other astronomers, together with the use of fairly accurate pendulum clocks, effected a refinement of astronomical observation from the four minutes of error of Tycho Brahe to the ten seconds of Flamsteed.[10]

One astronomer who stuck to the old ways was Johan Hevelius (1611–87) of Danzig, a private observer in friendly correspondence with the Royal Society. In 1647 he had published *Selenographia*, one of the most beautiful scientific books of the age, devoted to the surface features of the moon. Later his interest turned to the study of comets—phenomena of the skies which absorbed all astronomers at times. Hevelius strongly defended the old open-sighted instruments, engaging in a tedious controversy with Hooke which remained unsettled even after Edmond Halley had paid a visit to Danzig to compare the old and new types. Hevelius' conservatism rendered his posthumously published star-catalogue (1690) of little value. The finest achievement of the seventeenth century in mapping the stars was John Flamsteed's, whose *Historia Coelestis Britannica* appeared (also posthumously) in 1725. Flamsteed, as an ardent young astronomer, was taken up by various members of the Royal Society about 1670. Five years later he was chosen to be the first Astronomer-Royal. Although Greenwich Observatory was built by the Crown the Astronomer was left to equip it as best he might, so that Flamsteed laboured for many years with inadequate tools and in near-penury. His work was meticulous, in relation not only to the stars but to the sun, moon and planets. Tables of the moon's motion were important as a means of overcoming the longitude problem, and Newton was also in need of them to complete the complex gravitational theory of the moon's motion. Their correspondence, unfortunately, passed from civility to acrimony, especially when Flamsteed judged that Newton and the Royal Society of which he was then President had abused their rights with respect to the Observatory.

Edmond Halley, ten years junior to Flamsteed and his successor, the constant friend of Newton in this and other affairs, was without doubt the most gifted astronomer of his time if not the most exact. His interests in science were as varied as his talents. He put forward a new theory of terrestrial magnetism, drew the first chart of lines of equal magnetic variation, and suggested a connection between the Earth's field and the Aurora Borealis. He contributed several mathematical and physical papers to the *Philosophical Transactions*, which he also edited for a time when Secretary of the Royal Society. He assisted Newton in William III's recoinage, studied the antiquities of Roman Britain, read Arabic mathematics, and made chemical experiments. Halley was versatile in an age when most astronomers were tied to their constant watch of the night sky. He first attracted attention by an expedition to St Helena in 1676–8 to observe the southern stars, equipped with

an Excellent brass Sextant, of 5½ feet Radius, well fitted up, with Telescope Sights, indented Semicircles of the same metal, and screws for the ready bringing it into any place; A quadrant of about 2 foot Radius, which he chiefly intended for observations, to adjust his Clock ; a good pendulum Clock ; and a Telescope of 24 feet; some lesser ones; and two Micrometers.[11]

On this occasion Halley was transported in an East Indiaman. Later he made the first scientific surveys supported by the British Admiralty. But Halley was an able mathematician and his real status was as a theoretical astronomer. This explains why he brought Newton's *Principia* before the world (with his own ode of praise prefixed to it). His mathematical constructions mostly relate to technical problems of astronomy, especially the determination of orbits ; two of these are of wider interest. In the *Principia* Newton had demonstrated from the theory of gravitation that the orbit of a comet should be a conic, a parabola if the comet should not return, an ellipse if it could be shown that the same comet was seen more than once :

Comets [he said] are a kind of planets revolving in highly eccentric orbits about the sun. And just as the planets without tails are smaller when they revolve in smaller orbits and closer to the sun, so also it seems reasonable that the comets that approach the sun more closely at perihelion are smaller and revolve in smaller orbits. But I leave the conjugate diameters and periodic times of these orbits to be determined by comparing comets together which return in the same orbits after a long period of time.[12]

This was precisely what Halley did. In 1695 he satisfied himself that the motions of the comets of 1680 and 1682 fitted an elliptical path better than they did a parabolical one. The former he identified with the comet seen soon after the death of Julius Caesar, and others of A.D. 531 and 1106.[13] The comet of 1682 he identified with those of 1607 and 1531, and less certainly with those of 1305, 1380 and 1456. He correctly attributed disturbances in this recurrent comet's motion to the gravitational pull of Jupiter and hence was unable to predict the exact date of the next appearance in the winter of 1758–9. In fact, to the intense interest of astronomers, Halley's Comet reached perihelion on 13 March 1759, sixteen years after Halley's death and just a month earlier than the date calculated shortly before by the French mathematician Clairaut. The accuracy of Halley's prediction, as revised by Clairaut, was justly regarded as an outstanding triumph for the Newtonian theory of universal gravitation.

Halley's other major contribution to the strategy of astronomy was his programme, worked out in 1716, for taking full advantage of the transits of Venus across the face of the sun to occur in 1761 and 1769. The first European observations of planetary transits had only been made in the early seventeenth century, and their scientific possibilities had never yet been properly exploited. Halley carefully worked out the optimum procedure for determining from the transits of Venus the true value of the solar parallax, and hence the size of the solar system. As a consequence of the interest aroused by him the two transits became the occasion for the first major exercise in international scientific co-operation, and with fruitful results. In this matter as in others Halley acted as a far-sighted scientist, with a sense of the tasks of real importance ; yet, either for reasons of temperament or because of a lack of intellectual concentration he fell just short of making contributions of supreme power.

The ideas of the seventeenth century on the origin, history and present nature of the universe were wholly imaginative, for there were no facts on which to build. No better cosmological hypothesis succeeded that of Descartes and that was destroyed (if not at once discredited) by Newton's *Principia*. The next great steps were to come with Thomas Wright, Laplace and Herschel in the second half of the eighteenth century. The use of the telescope in the seventeenth century enabled astronomers to describe the solar system fairly accurately (though they missed both the very small planets or asteroids, and the very distant ones Uranus, Neptune and Pluto); thus the domestic question of cosmology was settled by the adoption of a modified "Copernican" system. After that emphasis shifted to celestial mechanics—to the problems solved by Newton in the *Principia*. But the telescope and even the *Principia* left the grand questions of cosmology, questions touching the multitudes of stars and the vast depths of space, utterly impregnable.

As immense size defeated astronomers, equally on the microscopic scale of things physical structure remained far beyond vision in its minuteness. Here there was no issue of such moment as the Earth's motion. Yet the microscope in its application to biology * opened up a wider and more various field of inquiry. It disclosed not a familiar world enlarged and brought nearer, but a wholly new world of unsuspected living forms, and—for the first time—the actual texture of living substance. It was possible for scientific astronomy to develop to the level of Newtonian theory, or very near it, without any use of the telescope at all ; scientific biology without any kind of microscope is almost inconceivable. That this

* The word biology was coined in the nineteenth century but its anachronistic use is too convenient to be set aside.

fact was not clearly seen in Newton's day—though the work of the three or four really first-class microscopists was generously appreciated—is easily explicable, for the invention of the telescope fell aptly just at the end of one tradition in astronomy, while the microscope was itself to be the basis of a new tradition in the study of living things.

For Galileo the telescope seemed to open up a new era in human knowledge. It was the weapon of evidence against which scholasticism had no defence. The microscope, on the other hand, through which appeared (as Galileo also said) a multitude of natural wonders, was still a toy : it made flies appear with hairs on their legs, as big as lambs. It was easy to regard the very small as a travesty of familiar, normal-sized objects, when even the smallest insect was found to possess its own intricate anatomy. Within a couple of decades of Galileo's death (1642) the microscope had become a more useful instrument than before, but it had still made much less effective progress than the telescope. If anything the limitations imposed by inadequate optical technique and theory were even more crippling in microscopy than in astronomy, and new biological techniques were slow to develop. There were fresh questions to be asked and no one yet knew how to formulate them. By about 1660 the compound microscope consisted of a biconvex objective, field lens, and single eye lens. The mirror for casting light upwards through a preparation into the lens-system was not to be invented for another thirty or forty years. This optical arrangement continued without essential change down to the beginning of the nineteenth century, the improvements of intervening years being almost wholly concerned with mechanical design. Indeed, in all this time much of the most useful work with the microscope was done with simple equipment and at low powers. What was important was to discover the type of task to which the limited technical means of the period could most usefully be applied.

The fact that a rather elaborate form of compound microscope with auxiliary equipment was described in Robert Hooke's *Micro-*

graphia (1665), and by other writers in the next few years, is not of serious significance. *Micrographia* was the first large book, with fine plates drawn by Christopher Wren, to be devoted to the new instrument ; but in it most of Hooke's description is at a superficial level and a fair part of the book is not about microscopy or even biology at all. Like his predecessors Hooke enlarged the flea, fly, ant and louse; more important, he saw in cork the dead walls of the plant cells (to which he first applied this term, from their resemblance to the cells of the honey-comb), he saw the plant-like form of mould and came near to discerning the nature of oolitic rock. Acute and varied as they were, however, Hooke's observations were far from realising the true biological significance of the microscope, for which simple means were quite adequate.

Its real usefulness began when it was appreciated that it could give answers to four different types of question : what is the structure of animal substance ? What is the material of which plants are made ? How do insects differ from animals, and both from plants ? Are there living organisms invisible to the naked eye ? None of these questions had been seriously considered in science before. In addition the microscope brought a new resource to old problems, like the development of egg and seed. Nearly all of this was connected more or less directly with the study of anatomy, and so in large measure the microscopic investigation of the seventeenth century is a specialised aspect of comparative anatomy. At the root here was one of the concepts of seventeenth-century science of greatest potential value, in biology as well as in physics, the idea that it is through examining the structure of things that an explanation of their way of functioning is discovered. Though there were differences in the applications of this concept to physics and biology, yet as some physicists looked to the theory of matter for the ultimate explanation of all physical phenomena, so through the microscope some biologists began to seek, thus early, in the organisation of living substance the principles of its vital activity. Once anatomy had found the essence of physiology in the arrangements of organs within the animal ; now the problem was to find how the organs performed their functions—and to discover more accurately what these functions were—from study of their minute structure.

The comparative interest in microscopy was strongest in the Dutch naturalist Jan Swammerdam (1637–80), who studied the anatomy and physiology of insects in just the same way that zoologists investigated those of animals. For this purpose he did not require high magnifications, but he did need very delicate dissecting-instruments and special techniques of preparation, which he devised. His researches began even before he took his medical degree at Leiden (1667) and absorbed him so greatly that he took little interest in his career as a physician. They were partly published in his Dutch book on insects (1669) and treatise on the may-fly (1675), written, he said, " to give us wretched mortals a lively image of the shortness of this present life, and thereby to induce us to aspire to a better ", but the bulk of his work was only printed in the *Bible of Nature* a century after his birth (1737). Swammerdam gave himself to natural history for about ten years only, for after 1673 he was increasingly preoccupied with harrowing religious contemplations and came to distrust science as a trivial distraction. The religion of a naturalist is often reflected in his pages :

> Whoever reads these things must be obliged to confess that the power of the Almighty cannot be known by clearer and more convincing proofs, than in those minute animalcules wherein that great Architect has inclosed and hidden so many wonderful parts, and shown such exquisite art, that exceeds all human industry ; so that one may employ his whole life in the dissection of the smallest of animals.[14]

In his last work, on the may-fly, Swammerdam gave the first complete account of metamorphosis, to his mind a universal stage in the development of living things. The finest study of insect anatomy he made is that of the bee, but he also dissected besides other insects the snail, cuttlefish, lobster, hermit-crab, several kinds of worm and the frog. At a time when belief in the spontaneous generation of the creatures he dissected was still common, Swammerdam provided the first clear description of their reproductive physiology, life cycle and development. He found that the snail is an hermaphrodite, that in the frog fertilisation occurs externally, that the queen-bee is the mother of the hive (not the king as Aristotle had supposed), the drones male, and the " females " neuter. He showed how different in the invertebrates is the pattern of living organisation from that familiar in the vertebrate phylum. He missed some points of detail—in the nervous system of the Ephemerid nymph, for example—but much in his descriptions was not superseded before the nineteenth century.

Though Swammerdam was by no means overlooked by his contemporaries—on a visit to Paris he repeated before a meeting at Thevenot's house his dissection of the may-fly, and an offer was made to tempt him to Florence—the microscopic anatomist who captured the imagination of the time was Marcello Malpighi (1628–94), for most of his life a Professor at Bologna, who was also earliest in the field. Two letters to Borelli on the structure of the lungs (in which the capillary circulation was first described), published in 1661, mark the first application of the microscope to the examination of living tissues. This was Malpighi's initial concern which he later extended to the brain, tongue, liver, kidney and spleen (1665, 1666). Minute anatomical structure was important to him as the key to understanding the function of the organ, though he could not pursue physiology far. The chemical aspects of physiological processes, of which he was well aware, were quite beyond his reach. He illustrated a taste-bud from the tongue and in the kidney described the " Malpighian capsule " where the tubules that secrete urine receive the fine blood-vessels. Later Malpighi embarked upon comparative anatomy with his superb monograph on the silkworm (1669), studied the development of the chick (1673, 1675) and plant anatomy as well (1672). Structure and its development are dominant themes throughout his work, all the latter part of which (after 1669) was sent to be read before the Royal Society and published in London. Malpighi touched on nearly every point of microscopic anatomy that was accessible to him, and the importance of his observations for understanding of function was supreme. In the silkworm book, for instance, he described for the first time how insects breathe through tracheae penetrating to each part of the body, and opening to the air in spiracles. Blocking these spiracles with oil caused loss of function in the region they served.

Very early in his investigations Malpighi also began to observe the structure of plants in the hope that their more simple organisation (as he thought) would serve as a guide to that of animals; conversely he was also tempted to construe what he saw in them by analogy with the functioning of animal organs. He described and beautifully illustrated the appearance of stems with their radial and longitudinal members, the yearly growth of the stem, the formation of the leaf and the connection of its veins with those of the stem, the growth of the young shoot from the seed, the anatomy of the inflorescence, and many other points, some of them also treated by his English contemporary Nehemiah Grew (1641–1712). The latter's drawings are more geometrically schematic and so less faithful than Malpighi's, nor did Grew use the method of sectioning to the same varied effect; on the other hand he wrote more on the chemical aspect of plant physiology—prematurely, as what he had to say was highly speculative.

The stamp of the professional anatomist on the work of Swammerdam and Malpighi—both had been trained in the medical schools—raises it to a wholly different level from that of the other early microscopists. Their work was the direct antecedent of the histological and cytological researches flourishing so brilliantly in the nineteenth century, and delayed until then not only by the imperfections of the instrument but by a decline of interest in problems of structure. The eighteenth-century microscopists were naturalists at heart, not anatomists. Some of the reason for this change lies in the extraordinary discoveries made by Antoni van Leeuwenhoek (1632–1723), whose strenuous period of activity as an observer fell precisely when that of Malpighi and Swammerdam was closing. Unlike them Leeuwenhoek was without higher education and without anatomical training, though he was familiar with the commonplaces of Cartesian science. When he began he did not know that tendons and nerves are distinct structures, and while attentive to the figures displayed by Hooke and other microscopists his ignorance of any language but Dutch prevented his reading their work. Everything he did was reported to the Royal Society, many of his letters being translated for the *Philosophical Transactions*: but one secret he steadfastly kept to himself was that of the manufacture and use of his extremely high-powered single lenses. Not even his most illustrious visitor was permitted to see the best of them. Like others, Leeuwenhoek examined animal and vegetable tissues, without remarkable results ; his curiosity also led him to take as specimens crystals, seeds, insects, stones, milk, blood (he was one of the first to see the red corpuscles) and all kinds of artificial materials. Crossing one of the meres near Leiden in the summer of 1674, when the water was made green by algae, he placed some of the greenish water before his lens, and so saw the free-living protozoa of pondwater. His letter (7 September 1674) seems to describe both ciliates and flagellates :

These animalcules had divers colours, some being whitish, others pellucid ; others had green and very shining little scales, others again were green in the middle, and before and behind white, others greyish. And the motion of most of them in the water was so swift, and so various upwards, downwards and round about, that I confess I could not but wonder at it. I judge that some of these little creatures were above a thousand times smaller than the smallest ones which I have hitherto seen in cheese, wheat flour, mould and the like.[15]

At the same time he saw some small multicellular organisms,

among them rotifers and (later) hydra. All these "little animals" were of course far more minute than the smallest living things hitherto known, like "vinegar eels" (nematode worms) or cheese-mites. In a letter of pamphlet size (9 October 1676) Leeuwenhoek described at length the many different kinds of organism he had seen in rain-water, well-water and infusions of various materials. The document is reckoned as the first in the history of protozoology, but unfortunately the writer had no greater skill as a draughtsman than as a linguist and the letter was unaccompanied by sketches. Later he observed several of the species of protozoa parasitic on man and and animals, even bacteria, though the latter he could not characterise very clearly. Furthermore Leeuwenhoek was the first to report, and continuously investigate, the spermatozoa of men and animals.[16] Wherever he turned his microscope, even to sea-water, the dust of his gutters baked by the summer sun, or infusions of pepper and spice, it seemed that all organic matter was pullulating with minuscule life. This discovery with the microscope was, inevitably, that which most fascinated the minds of Leeuwenhoek's contemporaries and successors, though they could make little more of it in general or interpretative terms than he did. While the anatomical and histological researches of Malpighi and Swammerdam were ad-mired but not emulated—at least after the large work of Réaumur in the early eighteenth century—the favourite specimen of later microscopists was a drop of pond-water. Over-emphasis on an exciting new field caused microscopy to become a rather narrow kind of research.

For the actual animalculae of pond-water were no less difficult to understand than the imaginary animals in the moon would have been. By sheer accident—or curiosity—Leeuwenhoek, chamberlain to the aldermen of Delft, had hit upon the frontier between uni-cellular and multicellular organisms. When he turned to micro-dissection, as he did on occasion, his reports on the anatomy of insects fell into the pattern of contemporary research, as did even something so unexpected as his discovery of parthenogenesis in aphids. Insects had organs, like animals, digested, bred and grew in a manner analogous to that of animals. But when he observed his " little animals " Leeuwenhoek—though he was totally unaware of the fact—was observing single living cells. And the concept of the cell, despite Hooke's use of the word, and other microscopists' description of structures like bladders making up vegetable tissue, was one that biology lacked. The smallest living things now revealed were in principle incomprehensible to naturalists, since every attempt to assimilate them to the metazoa was bound to be erroneous. Though no praise of Leeuwenhoek's skill and persistence can be too high, his discoveries were confused in their origin and brought nothing but confusion as their first-fruit into the study of living things. The confusion could only be sorted out by distinguishing the part from the whole, when the cell was recognised as the unit of biological architecture.

The low-power microscopy of the later seventeenth century, comparable in its role to astronomical observation of the solar system, was still in solid contact with the gross anatomy of men and animals and hence with the prevailing body of biological ideas, imperfect as these were. If effectively continued, it might have brought about a sounder development of physiological knowledge. High-power microscopy, like a telescope turned to the stars, faced insoluble problems. The ideas it produced were, like those of the cosmologists, wild speculations little related to the known facts of the limited, accessible scale of dimension. Merely to see the multiplied crowd of stars through a telescope was no more aid to understanding their nature than to see the multitude of living animalcules through a microscope was to understand the nature of living organisms. Optical instruments gave the seventeenth century a tantalising glimpse of the very large and the very small in nature, but the development of knowledge and ideas necessary for understanding at these unfamiliar levels could not be gained in a single step. Attempts to interpret hastily the vastly greater scale of things from the mean position of man were bound to be deceptive.

It was just as false to suppose that a spermatozoon has anatomical structure as to imagine that a star burns like the kitchen fire ; but no other patterns of explanation for either were yet available. To attain a truer type of explanation it was necessary to retreat, after this enticing glimpse of wonders, from the extremes to a more moderate order of size.

CHAPTER VII

PROBLEMS OF LIVING THINGS

Now to this Hydraulic Organ ye may compare a Beast, whose Soul being indeed, by reason of a certain modification of her matter, qualified to perceive the various impressions made by Objects upon the Nerves of the instruments of the Senses ; and to perform many trains of Action thereupon ; is yet so limited in her Energy, that she can perform no other actions but such as are (like the various parts of an harmonical Composition) regularly prescribed (as the Notes of a Tune are prict down on the tumbrell of our Instrument) by the Law of her Nature, and determined for the most part to the same scope, the Conservation of herself and the Body she animates. So that she seems qualified only to produce a Harmony of Life, Sense and Motion : and this only from a certain contexture of the spirituose Particles of the matter of which she is made, and from the respective Organization of the Body in which she acts.[1]

Nowhere are the Hellenistic antecedents of modern science more clearly apparent than in natural history and the medical sciences. The abrupt return in the sixteenth century to the biological writings of Aristotle, Theophrastos and Dioscorides, and to the medical books of Hippocrates, Galen and Celsus, is a paradigm case of the Renaissance mind at work. Except with respect to surgery— rather a craft than a science—what had happened between A.D. 200 and A.D. 1400 was practically cut out of biological knowledge. There were but three survivals from the Middle Ages: the university medical school; the practice of anatomical dissection; and the use of certain pharmacological preparations, distilled liquors chief among them, unknown to the ancient world. With these preserved little was sacrificed by beginning biology and medicine over again where the Greeks had left off.

No Duhem has appeared to forge chains of continuous evolution between medieval scholasticism and modern biological science which just as certainly, though far less confidently than modern physics, takes shape in the seventeenth century. Research among manuscripts by scholars from Sudhoff to Thorndike has shown that the sheer volume of biological, or more exactly medical, writing in the Middle Ages was larger than it was once assumed to be, and that there was not such utter neglect of straightforward descriptive science—anatomy and botany for instance—as was formerly supposed. But there is no hint from the Middle Ages as yet of anything that could transform understanding of the complex processes that constitute life, of anything parallel to the impetus theory and the kinematics of Oxford and Paris. The reason why nothing of this kind is likely to be found is obvious enough. Scholastic philosophy dealt with ideas ; and living processes could not be studied as ideas. There could be few ideas worthy of discussion in biology until a long and (to the outsider) often tedious course of investigation into the facts, and the mechanisms underlying the facts, had been run through. There is very little evidence that the Middle Ages initiated such investigation or saw the necessity for it.

Physics as a science has grown experimentally upon a skeleton of ideas, however tenuous and variable. That is why it has proved so fascinating to philosophers and intellectual historians; strangely enough men have been more articulate about their notion of the universe than about their notion of life. Biological ideas through most of history consisted of borrowed theology and commonplace metaphysics; they were often highly subjective and always dogmatic. In the eyes of those who regard science as a body of theory explaining the universe, rather than as a theoretically organised body of knowledge about the universe, the history of

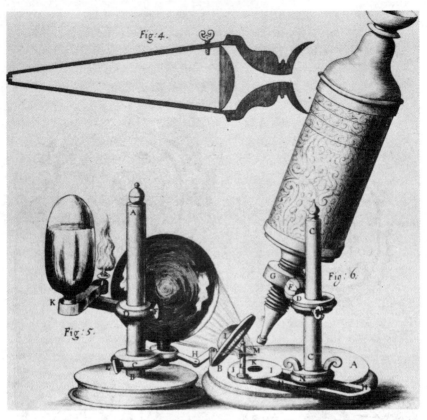

Hooke's microscope, 1665, with its lamp and bulls-eye, and (*below*) his drawings of the mite. From *Micrographia*

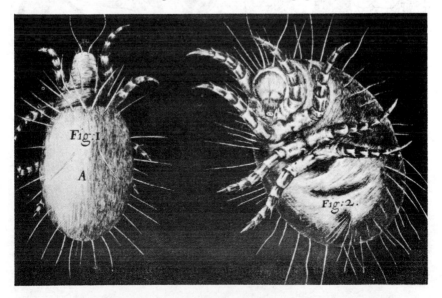

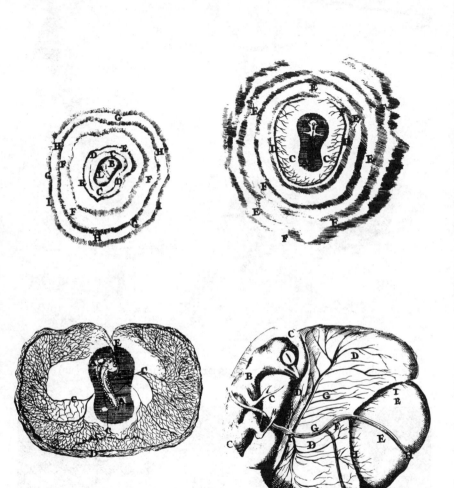

Four of Malpighi's drawings of the development of the chick embryo,
from the first signs of structure in the egg

biology until the nineteenth century lacks significant cohesion ; natural history was a merely Baconian compilation of facts, and medicine a fumbling, superstitious craft. In a measure this judgement is just, for only the collector's joy in acquisition can make collection other than a tedious labour. The great systematising and collecting naturalists, the anatomists, the microscopists, were only incidentally men of ideas ; for the most part they were content merely to know. And the point is, of course, that without this simple desire theoretical biology—expressed in the theory of evolution, genetics, chemical physiology—could never have existed.

For the grey, sober truth is that the only practicable first approach to the study of living things was the Baconian one of collection and comparison. Living processes are too varied in their manifestations and too complex in their chemical character to yield to *a priori* reasoning. The first requisite for physiological explanation is recognition of differences in type and structure—between flowering plants and cryptogams, vertebrates and invertebrates, animals with lungs and those without. Classification, which to be rational must be founded upon an objective analysis of the morphology of the organism, has as its purpose orderly comparison of species; comparison provokes the questions that experimental, theoretical science endeavours to answer.

Collection, description and classification were still the dominant themes of botany and zoology during the seventeenth century, as they were to be for another hundred years. Few intelligent men thought that this rather humble level of activity required justification, since they saw the world as there to be described, plants as well as planets. Natural history, strongly anthropocentric in the Renaissance, became more objective in its outlook and methods, considering neither plants solely for their medical utility nor animals for their approximation to humanity. As always in early science loss of the human focus caused some disintegration. How was knowledge of plants to be organised if usefulness to man was not the prime consideration ? The arrangement of one of the last herbals in the Renaissance tradition, John Parkinson's *Theatrum Botanicum* (1640), strikes a compromise. Six of his seventeen groups of plants are frankly subjective : into them went the plants that were sweet-smelling, purgative, opiate and poisonous, cooling, hot and sharp in taste, and fit for healing wounds. Seven other groups are roughly based on morphology : the umbelliferous plants, ferns, grasses and reeds, leguminous plants, cereals, thistles and other thorny plants, and trees and shrubs. Saxifrages or " Breakestone Plants " and a miscellany of " Marsh, Water and Sea Plants, and Mosses, and Mushrooms " are two groups each connected partly by morphology, partly by habitat. (The two groups required to complete the tally, " The Unordered Tribe " and " Strange and Outlandish Plants ", have no obvious coherence.) From this time forward subjective classification was abandoned in favour of entirely objective treatment, such as had been adumbrated by Cesalpino (1519–1603) in the sixteenth century, which was largely based on morphology, partly on habitat. The hunt for a truly " natural " classification—one that should relate each species to every other species by considering all of its characters, and not merely a few selected as suitable for rapid codification—was now on, nor did it cease even with the success of the " artificial ", mechanical taxonomy of Carl Linnaeus (1707–78) in the eighteenth century. The idea that when properly understood nature is always found to be consistent and logical was as common in biology as in the physical sciences. The concept of a natural classification is indeed a kind of biological Platonism, since it assumes that if only the taxonomist can seize upon the ideal form of the species lying before him he can infallibly assign it to an ideal pigeon-hole in the order of things.

Seventeenth-century botanists, needing already some 18,000 pigeon-holes for their plants, could not succeed in arranging them in perfectly symmetrical cabinets.[2] The difficulty was to choose those characteristics of structure that significantly differentiated one plant from another and in turn indicated that one group so formed was more closely allied to a second than to a third. It is easy to hit upon some obviously similar plants, like the conifers, but not all fall into obvious groups and the relations between groups are still more obscure. The great French botanist, Joseph Pitton de Tournefort (1656–1708), professor at the Paris Jardin des Plantes, Academician and an ardent collector, anticipated Linnaeus by classifying solely by flower and fruit. Root, leaf and stem he regarded as completely unreliable as criteria of relationship. In spite of this he missed the prime distinction between flowering and non-flowering plants. The centre of his taxonomy was the genus, hazily split into species and assembled with others of like kind into classes.

In England John Ray (1627–1705) took a more eclectic view. Though he began as a botanist, his first publication being a catalogue of the plants to be found near Cambridge, his interests broadened to include every aspect of natural history. After the loss of his Fellowship at Trinity (through his refusal to subscribe to the Restoration religious tests) he worked on birds and fishes with his patron Francis Willughby (1635–72), travelling extensively in England and on the continent. Ray studied the animals as well, and towards the end of his life tackled the insects. His great work, however, was *Historia Plantarum* in three massive folios (1686, 1688, 1704). He was well aware of the imperfections of taxonomy :

> I would not have my readers expect something complete or perfect [he wrote in the preface to *Methodus Plantarum*, 1682], something which would divide all plants so exactly as to include every species without leaving any in positions anomalous or peculiar ; something which would so define each genus by its own characteristics that no species would be left, so to speak, homeless or be found common to many genera. Nature does not permit anything of the sort. Nature, as the saying goes, makes no jumps and passes from extreme to extreme only through a mean. She always produces species intermediate between higher and lower types, species of doubtful classification linking one type with another and having something in

common with both—as for example the so-called zoophytes between plants and animals.[3]

In classifying plants he employed a difference among seeds that he regarded as " first and best of all ", that between plants whose embryo contains a single lobe or cotyledon, and those containing two. (Malpighi had already proved by experiment that the cotyledons supply nourishment to the developing plant before it acquires a root-system.) For trees and shrubs he freely adopted existing groupings, relying heavily—and mistakenly—on the form of the fruit. Dealing separately with algae, fungi and lichens, mosses, and herbs Ray distinguished the flowering plants by their flowers, seeds, fruits and leaves. He hit upon many of the natural families of modern taxonomy and avoided the irrelevant characters to which appeal had so often been made in the past. Though some of his groups appear hopelessly jumbled to a botanist of today the principles of his classification were as natural as he could make them ; where he went wrong it was often because he was ignorant of the true structure of the plant.

Botanists have sometimes, especially in the late eighteenth century, seemed to overlook the distinction between the purpose of classification—to trace the natural affinities between plants—and its usefulness in providing a code for identifying them. Ray did not, recognising that a code, however helpful to beginners learning names, had no real justification in nature. He was much interested in plant anatomy and physiology, making good use of the work of Malpighi and Grew though no microscopist himself. The principles derived from their analysis of plant form are infused into Ray's taxonomy. His insight into the importance of morphology is even easier to illustrate from his classification of animals, described eleven years after the *Methodus Plantarum* in *Synopsis methodica Animalium* (" Methodical Survey of Animals ") of 1693. His arrangement of sanguinous (that is, vertebrate) animals is summarised in figure 8.

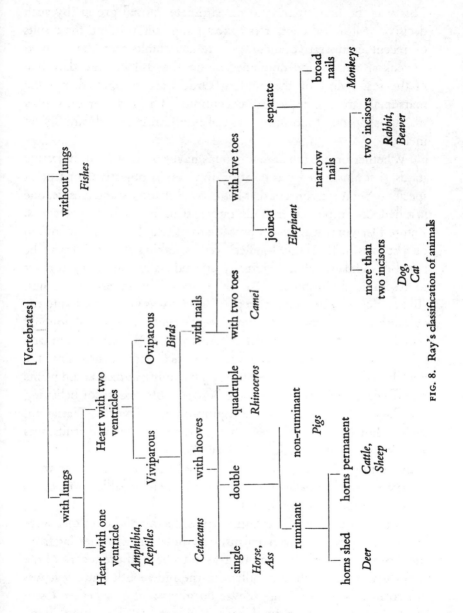

FIG. 8. Ray's classification of animals

This was the first attempt to discriminate animal groups by such decisive anatomical characters as teeth and feet, utilising the results of recent comparative anatomy ; it is notable that Ray refused to reckon whales and dolphins among the fish because they live in the sea. Most of the modern Orders are recognised, but the marsupials are a conspicuous exception. The first anatomy of a marsupial species (opossum) was only published by Edward Tyson in 1698.

Whether or not the higher groups have an other than subjective status, a sophisticated classification implies an objective concept of species. Yet it is a concept that has proved shadowy and elusive, one justified on empirical grounds rather than by definite principles. Before Darwin it was at least possible to offer a theoretical definition of a species in theological-genetic terms, taking its members to be lineal descendants of an originally created pair, continuing forever to resemble their antediluvian ancestors on every essential point. (In practice of course this concept of species was utterly useless to the taxonomist.) No one in the seventeenth century seriously doubted that extant species had originated in this way, and bred true. In fact, this belief provided the strongest theoretical argument against the doctrine of spontaneous generation. That animals, insects and plants reproduce their kind now was the best possible reason for believing that they have done so since the beginning of the world. Variation, such as horticulturalists encouraged in their blooms and fruits, was a pathological abnormality. As Ray said :

the number of species in nature is fixed and limited, and as we may reasonably believe, constant and unchangeable from the first creation to the present day.[4]

In that case what of the fossils, shells, teeth and bones that were discovered, often so inconveniently in mountain quarries far from the shoreline, where they had no reason to be ? If the record of the rocks was a troublesome problem in the nineteenth century it was not so in the seventeenth. It was no great matter whether fossils were thought to be natural rock accidentally mimicking living forms, or the actual relics of living organisms, for few supposed that the later were closely related to modern species. In most instances it was clear that fossils were very different from anything found alive in the same region. Each fossil-bed, the English naturalist Martin Lister (1638?–1712) pointed out (1671), displayed its own singular species, none resembling " the Spoils of the Sea, or Fresh-Waters, or the Land-Snails ". Hence he declared them to be true rock, observing that each fossil was of the same substance of its matrix, and not cast " in any animal mould, whose species or race is yet to be found in being at this day ".[5] In which he was in general right, but Lister meant that fossils were not of organic origin at all. Robert Plot, who gives some neat figures of fossil shells in his *Natural History of Oxfordshire* (1677), was of the same mind. It was for him too much to believe that sea-creatures had been where the sea never was, even during Noah's Flood. Others, like Leonardo da Vinci and Fracastoro, had thought otherwise many years before. Athanasius Kircher (1665) suggested that fossils were organic remains ; Robert Hooke, in *Micrographia* in the same year, thought they were formed like petrified wood :

all these do owe their formation and figuration not to any kind of Plastick virtue inherent in the earth, but to the shells of certain Shel-fishes, which by some Deluge came to be thrown to that place, and there to be filled with some kind of Mudd or Clay, or petrifying Water which in time has been settled together and hardned in those shelly moulds into those shaped substances we now find them.[6]

Steno in his *Prodromus* (1669) advocated a similar opinion, and stressed the close similarity between fossil sharks' teeth and those of the extant fish. Ray was hesitantly on the same side, while admitting that all objections to the idea of the organic origin of fossils could not be answered. For to adopt this idea was to suppose (with Steno) that the fossils were laid down when the whole Earth was covered by the Flood, or to commit oneself to some such hypothesis as Hooke's, that a shift in the Earth's axis of rotation had caused land once under the sea to become dry. The one thing that discussion of fossils did not do was cast doubt on the immutability of species (or of course the Biblical narrative). If ammonites a foot across had once lived their descendants had vanished ; yet no one was quite prepared to believe that God had created species and then killed them off in the Flood, and the strength of the contention that fossils are of organic origin rested on the view that fossils do closely resemble living species. Palaeontology could only become a significant science when the geological time-scale was vastly extended.

Of such an extension there was as yet no hint. Newton, Ray and everyone else was content to believe that the world was created less than six thousand years before. This was a revealed truth, even if it made it difficult to understand how in so brief a time mountains could have been raised by volcanic action, a channel cut by the sea between England and France, or great plains like the Netherlands formed by the deposition of river mud. Yet it was clear that the Earth had changed in such ways, without trace in the ancient memory of man. No theory of its formation, whether like those of Descartes and Leibniz assuming the crusting over of an incandescent fluid globe, or like those of Burnet and Woodward mainly directed to making the Flood an event of natural history, could escape the well-known differences in rock-structure or the colossal distortions suffered by their strata. The distinction between igneous and sedimentary rock was already made, and Steno in 1669 formulated the principle that all tilted rock was once horizontal. He speculated on the ways in which the tilting of strata could be caused, sometimes with a second layer of deposition occurring after the tilt. Here, it might seem, science cried out for a Galileo. In fact geology did not require theories—there were far too many of them. What it needed was measurement, measurement to determine how long was required to lay down a bed of limestone. It was not done and time stood still.

Even a century after the *Origin of Species* there are those who allege (ignorantly) that there is no good direct evidence for the mutability of extant species. In the seventeenth century cases of the adaptation of a species to its environment, and of more striking mutations, simply were not recognised, partly because taxonomy and field-studies were not intensive enough to make them appear. Rather the opposite occurred, for comparative-anatomical and other biological studies tended to bring in all the time fresh evidence of the uniqueness of each species, and especially of the mechanisms by which each reproduces itself. There was clearly nothing casual or indefinite about the machinery by which life is continued. Who would have expected to find the curious doubling of the sexual organs in the rattlesnake and opossum discovered by Edward Tyson (1683, 1698), the hermaphroditic mating of the snail, and the peculiar pads on the hands of the frog described by Swammerdam, whose only function was to grip the female in the sexual act? Or the strange scent-gland on the back of the peccary, noted by Tyson as the greatest wonder of the species, that "differences it from any other animal I know of in the world"? Detailed observation showed over and over again the marked distinctions between one species and another. Perhaps study of the cetaceans might have given an inkling of the effects wrought by adaptive evolution upon morphology. Tyson, in his anatomy of the porpoise, like Ray and others earlier, observed the difference of its fins from those of fish, the resemblance of its brain and viscera to those of land-mammals, the *rete mirabile* like that of ruminants, and he even described the pelvic bones. Yet in the end Tyson hesitated to declare that the porpoise is not a fish (being less bold than some of his predecessors) and no one yet recognised in these mammals of the sea animals whose basic plan had altered during countless generations to fit them for their aquatic habits. They had been created so, as one of the many apparent anomalies of nature, unique in the sea as the bat in the air.

The last breach in the tight inclusiveness of the concept of species was closed when spontaneous generation was discredited. It was a logically necessary step, since few really believed any longer that living forms could spring from dead matter or from any other source than a " seed " originating from parents of the same kind. What had been a philosophical theory in Aristotle's biology had degenerated into a vulgar, superstitious anachronism. Harvey was the last naturalist to give it any serious countenance. When experimental disproof of spontaneous generation was furnished by Francesco Redi (1626–98) in 1668 it carried immediate conviction. Shielded from flies, meat putrefied without producing maggots, while the maggots crawling on " control " pieces exposed to the flies developed into other flies exactly like their parents. Indeed, studies of insect anatomy—Redi also dissected flies to find their ovaries and oviducts—made the experiments almost superfluous. No one could yet, however, account for the origin of intestinal parasites ; that their eggs taken in food could " endure the heat of the intestines and the fluids passing through them ", said Swammerdam, was " a thing no man in his senses can admit ".[7] Indeed, since the atmosphere was so hostile the case for spontaneous generation was practically lost by default. It was simple enough to prove that worms, frogs and insects had parents, but the same had by no means been demonstrated of the " little animals " discovered by Leeuwenhoek. That problem was postponed to the next century. For the seventeenth the detachment of science from superstition required that spontaneous generation be impossible, just as in the twentieth (in very different setting) it requires the opposite.

Though the age of Newton and Ray took it for certain that each generation is virtually the replica of its parents it had no inkling of any method by which such replication was assured. Lacking all idea of genetic inheritance it developed very strongly the notion that one parent—either male or female according to which version was preferred—developed a complete, invisibly minute embryo ; the essential function of fertilisation was to trigger off the swelling enlargement of this embryo to visible proportions and ultimately independent existence. The embryonic form was not caused to take shape in the previously inchoate, unshaped matter of an egg or seed, as in the epigenetic theory taught (from Aristotle) by William Harvey in his *Generation of Animals* (1651) ; it was complete and preformed in all its parts before ever any development began. There was thus no need for a plastic virtue, or principle, or living force to sculpture the unorganised egg into resemblance of its parents. Rather the form and the potentiality for growth came directly from one parent in much the same way that the form and growth of hair, nails and skin (or in some creatures the regeneration of lost members) occur in the adult. It was mysterious enough that a lobster could regenerate a claw exactly to replace a severed one, but by the preformation theory the placing of an embryo chick in its egg by the hen could be no more mysterious than that ; whereas an egg unorganised, yet capable somehow of shaping itself into a chick, seemed a great deal more strange. Alternatively, in the extreme type of preformation theory, it was supposed that the embryonic seeds of every generation encapsulated one within the other, and had all existed from the moment of the first creation so that the embryo was never formed, for it had always existed. Eve was then literally the mother of the human race (or Adam its father). Here again and very conspicuously the notion of specific variation was not so much excluded as totally ignored.

While the theory of preformation surmounted some difficulties it could not explain clearly how in sexual reproduction the non-embryo-forming parent could leave imprints on the offspring, a problem Aristotle had faced two thousand years before. How could a maternally preformed embryo know whether its father would have blue eyes or brown ? And what of hybrids like the mule ? There are perhaps two reasons for the utter neglect by seventeenth-century writers of the problem of inheritance. The first is that much of the argument for preformation was drawn from entomology and botany. It is true that, when the preformation theory was really ascendant (after 1679), the sexual reproduction of insects and plants was known. But, as far as plants were concerned, the version of sexual reproduction adopted by Nehemiah Grew was hermaphroditic—each plant fertilised itself. The notion that separate plants fertilise each other was only put forward by Camerarius in 1696. And in insects minor variations attributable to parental influence are inconspicuous. The second reason is to be found in the link between the preformation concept and microscopic observation of the developing egg and seed. Focussing attention on this problem, where again parental variations were almost indetectable, the preformationists never compared their conclusions at this level with the familiar (and significant) fact that offspring do not all exactly resemble one parent. In other words in trying to account for the gross replication and continuity of species by assuming preformation of the embryo the more subtle problem of individual variety escaped out of sight. As indeed it did with virtually all naturalists, preformationists or not, down to the time of Lamarck.

Preformation theory was gaining ground before ever the microscope was directed to egg and seed. It sprang from a naive desire to prefer the simpler of two hypotheses at whatever cost in probability, though it was only given its full force and effect by the new instrument, with which naturalists soon found what they sought. The first English writer on the microscope, Henry Power (1623–68), consequently affirmed roundly (1664) " that every part within us is entirely made, when the whole organ seems too little to have any parts at all ".[8] For since the smallest living things then known were seen to have parts it was fair enough to judge that the smallest embryo, or the protozoa of Leeuwenhoek later, must have a complete structure of internal organs too. The abstract concept of infinite smallness laid an easy trap for those who held that organic structure (like atomic structure) could far exceed the possible limits of vision, or even imagination. Swammerdam sprang this trap on contemporaries when he set the entomological basis for the preformation doctrine in his book on insects :

The nymph or crysalis [he wrote] may be considered as the winged animal itself hid under this particular form. Whence it follows that in reality the Caterpillar or worm is not changed into a Nymph or Crysalis; nor, to go a step further, the Nymph or Crysalis into a winged creature. But the same worm or Caterpillar which by casting its skin assumes the form of a Nymph, becomes afterwards a winged creature. [This change is the same as that] observed in chicks [developing] from eggs, which are not transformed into cocks and hens, but grow to be such by expansion of parts already formed. Similarly the Tadpole is not changed into a Frog, but becomes a Frog by an unfolding and increasing of some of its parts.[9]

In Swammerdam's eyes one of the chief merits of the preformation theory was that it enabled him to set the embryonic stages of all living things—insect, frog, mammal—into the same pattern of development. Just as the tadpole came from a worm in an egg and turned into a frog, so man like all creatures began life as a maggot enclosed in an egg; a maggot which, wrapped in the amnion, underwent a nymph-like metamorphosis from which the embryo emerged in adult shape.[10] This was the first hint of the biological hypothesis of recapitulation. Yet Swammerdam's record of his observations continued undistorted by his theories. He admitted (without comment) his inability to discern a tiny embryo frog in the black granules of the frog's egg; he could also in speaking of the human embryo write of the limbs sprouting from the shoulders and lower parts of the body, at first like flower-buds, then growing out and dividing into jointed members.

Malpighi, as Cole has pointed out, was no more consistent. His series of figures, accompanying two tracts on the development of the chick sent to the Royal Society in 1672, are admirable within their range. Opening eggs at successively longer intervals after incubation began, Malpighi clearly sketched actual changes in structure and not mere enlargement of the parts, and himself spoke of seeing *rudimentary* organs, such as the liver and the eye. Obviously from these figures the embryonic chick was not always present, perfectly formed, waiting to be swollen to due size. Yet Malpighi held that the brain, the spinal cord, the heart and other organs of the chick were present before the embryo moved or showed red blood, before the egg was even laid. The parts of the chick pre-existed, like Japanese flowers, in the invisible germ of the egg, as did the plant in its seed.

The situation is easily paralleled in pre-Galilean physical science. Indeed, the anomaly was openly confessed by the later preformationists ; in Charles Bonnet's words (1762) :

All that I have just expounded on generation you may regard if you wish as nothing but a romance. I am myself strongly disposed to look at it in that way. I feel that I have only imperfectly explained the phenomena. But I ask you, can you find another hypothesis that will explain them better ?[11]

Like the odder passages in Newton's *Principia* the preformation theory of generation offers admittance to the back-door of seventeenth- and eighteenth-century science. The positive assertion of its truth by a host of competent observers, many of them fully acquainted with its manifest implausibilities, was a violent contradiction of their overt attachment to positivist or rational science. Preformation theory exemplifies the weaknesses of the mode of thought that was so eminently successful in the physical sciences. In their denial that unorganised matter could become organised—strictly parallel to the denial that dead matter could ever spontaneously generate living—the preformationists with their distaste for plastic virtues and vital principles sought to adapt mechanistic thinking to a biological problem. But on the other hand, like the physicists who deplored Epicurean atheism, biologists could not believe that the evolution of an organism from an egg or seed could be a completely mechanical process. Only the enlargement of living structures could be mechanical, not the first inception that determined their pattern. And so, as the mechanistic analogy between the universe and a great clock nevertheless demanded God the clockmaker to create the universe, so in biology mechanistic speculation demanded an origin for living forms that was not mechanical. If the animal was a machine (as Descartes held) nevertheless it had to be formed complete in the beginning as a little machine, since no machine could create itself out of inchoate matter. Of these notions emboîtement—the origination of all organisms not merely potentially but actually in the first and only creation of the world—was the logical and inevitable consequence.

At first the only known site for the preformed embryo in the reproduction of animals and insects was the egg. Harvey's faith that all living things arise from eggs seemed to be confirmed both by the entomologists and by Renier de Graaf's announcement, in 1672, that he had observed the eggs of viviparous creatures. De Graaf (1641-73) was a skilful anatomist, though he made no use of the microscope. The globular bodies of the size of a small pea that he saw in the Fallopian tubes of cow and rabbit were, however, the follicles or capsules in which the true egg is contained in the ovary. De Graaf never saw the true mammalian egg, one-tenth of an inch in diameter or less, which was first described by Karl von Baer in 1827. Despite his mistake the interpretation he gave was correct : he concluded that the organs of female mammals previously regarded as redundant testes were actually sources of ova that moved down the Fallopian tubes to develop, if fertilised, in the wall of the uterus.[12] Fertilisation he thought to be effected by a volatile spirit given off from the semen of the male.

Leeuwenhoek attacked this theory vigorously from the moment of writing his first letter describing the discovery of spermatozoa.[13] He denied that the bodies found by de Graaf were eggs. Rather, he maintained, the embryo originated in the male :

If your Harvey and our de Graaf had seen the hundredth part they would have stated as I did that it is exclusively the male semen that forms the foetus, and that all the woman may contribute only serves to receive the semen and feed it.

And at first Leeuwenhoek claimed to see rudimentary vessels and organs in the semen, not in the spermatozoa. Later he changed his mind and declared that the embryo was held within the sperm (of which he said he could recognise two types, male and female) ; in mammals the sperm-embryo found a niche to develop in the uterus whereas in egg-laying creatures it had to penetrate to the germinal spot in the yolk of the egg.[14]   With much bad evidence on both sides contemporary opinion favoured de Graaf and ovist embryology, often dismissing Leeuwenhoek's spermatozoa as parasites, perhaps products of disease, no more significant than the eels in vinegar or the little animals of pond-water.   Indeed, the necessity of spermatozoa for conception remained doubtful until 1824 and the entry of the sperm into the egg was first observed only in 1875.

The theory of emboîtement might have been developed as an embryological version of mechanistic biology.   In fact it was not. No one was as yet willing to assert dogmatically that the transmission of life from one generation to the next is a purely mechanical process, though this would seem to have been an inescapable corollary of the Cartesian concept of the animal as a machine.   Views on reproduction remained somewhat anomalous, at a time when explanations of many physiological processes, less closely linked with the delicate question of the origin of life, were freely fitted to the mechanistic pattern.   Yet the anomaly is less striking than it appears.   For it was possible to account mechanistically for many particular details while preserving that belief in the singularity of the organism as a whole which is more appropriate to vitalism.   Despite the authority of Descartes the mechanism of seventeenth-century biology was piecemeal and subordinate, always overshadowed by consciousness of the divine plan and the divine endowment of life.   Harvey had set the model in his book on the circulation of the blood (1628). He had shown that everything fell into place when the question was treated in the manner of hydraulic engineering, but he had gained this triumph at the cost of thrusting aside all consideration of the function of the circulation and of its passage through the heart and lungs of animals.   Inevitably it was this automatic action of the heart in pumping the blood that Descartes seized upon—and distorted—as illustrating his concept of the body as a complex machine. Inevitably too in going beyond Harvey to depict the heart as a mere piece of a machine Descartes displayed the weakness of incipiently mechanistic biology : its utter lack of subtlety. Descartes' task was like that of explaining the working of radio in terms of a Watt steam-engine.

This apart, Descartes was not concerned with organic physiology where at this stage he could discover little to aid his purpose. Leaving aside the animal, the mere machine, for the present he addressed himself to the composition of that more complicated being, man, in whom he found a soul united to a machine-like animal body (*Traité de l'Homme*, 1662).[15] His physiology was founded upon this duality, which involved the interplay of freedom (the control of the body by the independent soul) with necessity (the mechanistic functioning of the body itself). As a result Descartes was the first to perceive a clear distinction between voluntary and involuntary physiological processes, obscurely noted since antiquity. According to Descartes the soul was situated in the pineal gland, at the precise centre of the human nervous system, near the base of the brain. This gland was the chief reservoir of the nervous fluid whose flow it directed under the commands of the soul. The idea is best expressed in Descartes' own revealing analogy drawn from the play of fountains in an aristocratic garden :

And indeed one may very well liken the nerves of the [animal] machine I have described to the pipes of the machines of these fountains ; its muscles and its tendons to the other different engines and springs that serve to move them ; and its animal spirits, of which the heart is the source and the ventricles of the brain the reservoirs, to the water that moves these engines. Moreover, respiration and other similar functions which are usual and natural in the [animal] machine and which depend on the flow of spirits are like the movements of a clock or of a mill, which the ordinary flow of water can make continuous. External objects, which by their mere presence impinge upon its organs of sensation and by this means cause it to move in various different ways, according to the manner in which the parts of the machine's brain are arranged, are like strangers who, entering into one of the grottoes among the fountains, are themselves the unwitting causes of the movements that take place before them. For they cannot enter there without treading upon plates so arranged that if, for example, they approach a bathing Diana they cause her to take cover among the bulrushes and if they go forward to follow her they cause a Neptune to advance upon them threatening with his trident. Or if they walk in another direction they cause a marine monster to come forth who spews water in their faces, or other things of the same sort according to the fancy of the engineers who have made them. And lastly when the rational soul resides in this machine, it has its principal seat in the brain, and is there like the fountaineer who must be by the reservoir whence issue all the pipes of the machines when he wishes to set them going, or stop them, or in any way alter the movements they display.[16]

By virtue of its control over the flow of nervous fluid, Descartes explained, the soul determined all muscular motions, such as those of speech, giving effect to thought or the decisions of the will. For the nerves leading from the brain were fine tubes ending in the muscles ; when the fluid flowed along a tube the related muscle contracted, so moving an attached organ such as the eyeball or the larynx. But not all the motions of the body were directed in this manner by the soul. The peristaltic motion of the intestines clearly was not, and again in the case of the alternate relaxation and contraction of the thoracic muscles required for breathing Descartes described a beautifully ingenious mechanism whereby the nervous fluid was automatically led hither and thither to bring about the proper effect. In fact the human body was largely and the animal wholly self-regulating. Even its responses to external stimuli were mainly automatic. In the tubular nerves, said Descartes, were threads (fibres of marrow) conducted to the brain so that when a stimulus affected a sense-organ the threads were set in motion, opening valves in the brain and releasing nervous fluid to the appropriate muscles. In Descartes' example, if fire burns the foot a thread is actuated controlling the valves which (without demand from the soul) cause retraction of the foot, ejaculation of a cry of pain, and direction of the eyes to its source. Though his pathways were not those of a modern neurologist, Descartes clearly conceived of reflex action.

Indeed, the animal body was for him nothing but a bundle of reflexes, no more than a machine capable of responding spontaneously to stimuli from both its internal and external environment. Digestion, heart-beat, growth, waking and sleeping, sensation, formation of ideas, memories and emotions,

> all these functions [wrote Descartes] follow naturally in this Machine from the mere arrangement of its organs, neither more nor less than do the movements of a clock or automaton from that of its wheels and weights. So that there is no reason at all, on this account, to imagine in it any vegetative or sensitive soul, nor any other principle of movement and of life, than its blood and spirits excited by the heat of the fire that burns continually in its heart, and which does not differ in kind from all the fires which burn in inanimate bodies.[17]

Lacking a soul (and as Descartes believed a pineal gland in which to contain it) the animal was distinguished from other machines by its possession of a communications-system, its nerves. Only thought (with such movements of the body as sprang from conscious thought) revealed possession of a rational soul; so that the capability of thinking was that alone which differentiated men from machine-like brutes though their bodies were the same. In the *Discourse on Method* Descartes had suggested that a machine could be imagined so perfectly resembling an ape or other irrational beast that distinction between the two would be impossible. A man he thought could not be thus imitated in a mechanical simulacrum, firstly because of man's power of reasoned speech, since no machine could articulate sounds " so variously as to reply fittingly to what is said in its presence, as men of the lowest grade of intellect can do " ; and secondly because of the unique human ability " to act in all the occurrences of life, in the way in which our reason enables us to act ". Pressed now by the logic of his own position to determine in what man is not a mere machine, he defined the uniqueness of the Cartesian soul as its power of response to unprecedented situations : the responses of an animal, like those that could be built into a machine were (Descartes supposed) finite.

Whatever else it did—and of course Descartes' reasoning did not solve the problem of matter-spirit dualism—Cartesian physiology ended the contradiction between *mechanical* and *living*. A dead animal or plant was merely a worn-out or broken watch ; life was functional co-ordination of parts, not endowment with a vital principle or soul. The frontier was now drawn not between what is mechanical and what is alive, but between what is both living and thinking and what lives, but does not think. All below this line was directly and completely susceptible of objective scientific analysis. Only the analyst, the intellect, the ego that knows it exists because it thinks, remained impregnable to its own analysis. And it remained only to discover the real equivalents of the mechanisms—tubes, valves, threads and so on—that Descartes had postulated (for he had certainly not seen them in his dissections) in the bodies of animals and men.

As a philosophical or methodological principle this concept of Descartes is the most fecund in the history of biology. Three hundred years have added continually new revelations of its power. During the greater part of that time it was denounced by naturalists and biologists undoubtedly far more familiar with the details of the structure and behaviour of living things than Descartes was ; partly because it was " philosophical " (that is, not an induction from facts), partly because Descartes could only express his concept in terms of a very crude pulleys-and-wires automaton. Vitalism has proclaimed of every mechanistic " model " of living processes that it is far too clumsy to work ; that life is more complicated than physics and chemistry, and from each such stand it has been forced to retreat. True, every competent anatomist could see that Descartes' tubes and threads were fictitious, but the point is rather that he believed in the possibility of discovering the physical mechanism of nervous telegraphy. What matters is not the complete inaccuracy of both the anatomy and the physiology outlined by Descartes, but his tracing of the path along which vitalism would retreat as physics and chemistry became able to solve progressively more complex problems of physiological function. These problems were not hopelessly and as a matter of principle beyond human wit, since their solution was to be anticipated by the use of the very methods proving so fruitful in the physical sciences. Gravity, Galileo had said, was merely a name that explained nothing. Now Descartes declared that for biological studies life, too, was merely a name ; to invoke it as an explanation explained nothing.

For the physiologists of the late seventeenth century Descartes' writings were far more significant as a source of a method for trying to understand living processes than as the source of a biological philosophy. The beast-machine had a longer currency in the speculations of philosophers than in science, because working naturalists and anatomists never entertained it. They were at one with Descartes in believing that particular functions of the animal could be explained mechanistically without also holding as a principle that the whole animal was a machine. When they turned to the physics of living things they did so from a limited, rather than a total, faith in the mechanistic approach, nor did their impulse derive solely from Descartes ; it came from Galileo too (who had touched on biological statics in the *Discourses* of 1638) and from Pierre Gassendi. Their main focus of interest was an old problem—considered by Harvey without great profit—the nature of muscular action. The author of the principal book on this topic,

De motu animalium ("On the motion of Animals", 1680–1), was Giovanni Alfonso Borelli (1608–79), a mathematician and astronomer who also has a worthy place in the history of the theory of gravitation. His life was divided between Messina and Tuscany, where he worked with the disciples of Galileo in the Accademia del Cimento. Applying to nature as seen by the anatomist Galileo's phrase that its book is written in geometrical characters, he set himself to study the action of muscles as a branch of physico-mathematical science, and practised dissection for this purpose alone. In the early part of the book Borelli computed the effective pulls exerted by many different muscles, considering the leverage of the attachments of the muscles to the bones and the articulations of the joints (figure 9). Then, paying special attention to the location of the body's centre of gravity, he analysed the muscular motions involved in walking, running, lifting and so on, as well as the flight of birds and insects and the swimming of fish. All these he handled geometrically as demonstrations in applied mechanics.

In describing the contractile action of any individual muscle, Borelli followed the views put forward some years before, in 1664, by the Danish anatomist Steno.[18] Traditionally muscle-action had been ascribed to a contractile power in the fibres (uniting into the tendons at each end of the muscle), the fleshy mass being regarded as inactive. Vesalius, however, had asserted (1543) that the "flesh of muscles, which is different from everything else in the whole body, is the chief agent by which the muscle becomes thicker, shortens, and so moves the part to which it is attached", and in this opinion he seems to have been followed by Harvey, though not by all later anatomists.[19] Steno examined muscle microscopically, and concluded that it was the middle, fleshy portion of each fasciculus—made up of minute fibres—that did the work, becoming harder, shorter and more corrugated. He accounted for this contraction in his own mechanical terms, which Borelli rejected. Borelli had observed that when the volume of the ventricles of the heart diminished in systole the heart as a whole did not lessen in size. Therefore, he thought,

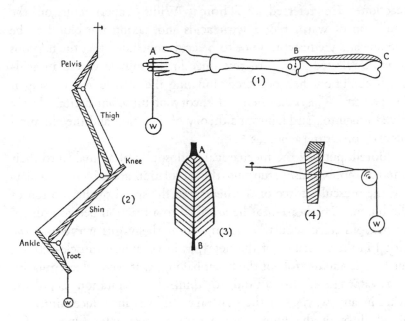

FIG. 9. Borelli's static analysis of muscles. (1) If w = 28 lbs., the pull sustained by the arm-muscle through the small lever OB = $\frac{OA}{20}$ is 560 lbs. (2) Model of forces in muscles of leg. (3) Though the muscle-fibres do not act so efficiently in an oblique as in a direct pull, the number of fibres is much increased; A, B, tendons. (4) Contraction of strings when a wedge is inserted

the fleshy mass of the heart must have swelled out into the ventricular spaces; when a muscle contracted it must actually increase in volume, as though the muscle-fibres shortened because something was forced between them, like a pair of strings contracting when a wedge is pushed through them. (Borelli took no notice of Steno's correct demonstration that such an increase in bulk cannot be measured.) Now the substance thus inflating the contracting muscle could not be pumped into by the nerves since these could not supply so much fluid, so Borelli guessed it came from a chemical reaction. He referred to Thomas Willis' experiments on the ebullition of warm blood with acids and, taking the blood to be alkaline and the nervous juice to be acid, concluded that the nervous impulse released a drop of this juice in the muscle where an acid-alkali ebullition then occurred, inflating the muscle and causing it to shorten. This first chemical theory of muscular action clearly owed much to van Helmont's theory of digestion and the chemical theories of Otto Tachenius.[20]

Borelli pursued the mechanical philosophy of animals into their internal motions also, studying the circulation attentively, calculating the muscular force of the heart and the speed of the motion of the blood. The heart-beat he ascribed to a reaction between drops of nervous juice regularly oozing from the vagus nerve and the blood in the substance of the heart, as in voluntary muscle-action ; the systole and diastole of the heart having, as it were, the regularity of a water-clock. Borelli did not think that respiration served the heart in any way. On the contrary, air was introduced into the blood through the lungs to perform a most vital function for the whole organism, but not one of a chemical kind. The purpose of respiration was purely mechanical : Borelli imagined that the elastic particles of air in the blood-stream, being compressed, vibrated like springs and so controlled the various periodic functions of the body, preventing it working either too quickly or too slowly. It is very interesting that Borelli realised the need for some control of the rate at which physiological processes occur, comparing the vibrations of the air-particles in the blood to those of a pendulum that regulates the going of a clock. Thus, he said,

> the automaton has a certain shadowy resemblance to animals, in so far as both are self-moving organic bodies which employ the laws of mechanics and are moved by natural powers.[21]

Borelli went on to explain many other functions of the body in these mechanical terms, but already he had deserted his strict geometrical theorems and the rest of his treatise is hardly less fanciful than is Descartes' account of the beast-machine. Like earlier (and later) naive mechanists Borelli sought to explain the action of the liver and kidneys as simple filtration, thought that the duty of the lungs was to thoroughly incorporate food-particles received from the thoracic duct with the blood, and so on. The idea that processes of chemical fermentation go on in the body, and the notion that the air brings in a vital nitrous material, he regarded as quite absurd.

Borelli's appeal to a chemical reaction in the muscles was thus practically unique in his theory. There were other speculative physiologists, however, who developed the concept of chemical mechanism in the living organism far more thoroughly. In the late seventeenth century this kind of explanation was characteristically English, though its roots lay much further back in the Germanic alchemical tradition. It was no novelty to suppose that there is an analogy between chemical and physiological processes : Galen had compared the digestive action of the liver to vinous fermentation and a common necessity for air had suggested to him a likeness between respiration and combustion. But Paracelsus (c. 1493–1541) was apparently the first to proclaim the identity of chemical and physiological digestion, in a nebulous fashion depending on his belief that all natural phenomena were in some mysterious way of chemical or alchemical origin. For Paracelsus this bodily chemistry was the opposite of mechanistic, since he postulated that it was managed by a spirit (archeus) residing in the stomach. Physiological chemistry could only play a useful part in science when objective chemical knowledge had been distinguished from alchemical mysticism and when it was possible to suggest some definite chemical process that was actually happening in the animal body.

Such a suggestion, though not yet divorced from esoteric overtones, was made by Johann Baptista van Helmont (1580–1648), whose ideas became widely known from about 1640 and especially after the publication of his collected works in 1648.[22] Van Helmont was by no means a mechanist, hardly even a " chemist " in the restricted, practical sense this word was now acquiring. In a philosophy that was far-ranging and vague to the point of incomprehensibility van Helmont took an animistic or organismic view of the universe. His ideas on the physiology of digestion, though more strictly chemical than others he entertained on living processes (regulated by the undefinable entity " Blas " or the " Spirit of Man "), were set in a context conforming little to the rationalist mood of later seventeenth-century science. He supposed that the first digestion, in the stomach, was accomplished by an acid " ferment " derived from the spleen, " the which failing, the appetite goes to ruin ". But the acidity of the stomach content would be harmful if it spread to the rest of the body, and so in the duodenum by a second " ferment " coming from the gall-bladder this " sharp volatile salt is changed into a salt volatile one ". This neutralisation he likened to the reaction whereby sharp vinegar becomes, with lead, the sweet-tasting *succerum Saturni* (lead acetate), and that in which any acid is " drunk up in an alcali salt ".[23] The neutral or alkaline food-material then passed through the mesenteric veins to the liver where it was converted into crude blood, for the rest of van Helmont's account followed an essentially Galenic pattern and took no heed of Aselli's recent discovery of the lacteal vessels ; * except that van Helmont wrote of unspecified " ferments " where Galen had attributed the formation of blood to the natural faculties of the organs and vessels.

To van Helmont the spontaneous changes taking place in organic bodies—whether the fermentation of the grape or the digestion of food—were still quite dissimilar from chemical changes *in vitro*. Acid-alkali digestion was more than a chemical process, he insisted,

* The importance of Aselli's observation only became clear when Jean Pecquet described (1651) the *receptaculum chyli*, the thoracic duct, and the junction of the latter with the vena cava. It was then possible to see that food did not pass through the mesenteric veins and the liver into the blood, as Galen and everyone else had supposed, but after being absorbed from the intestines by the lacteals issued through the thoracic duct directly into the blood-stream. Richard Lower (1669) performed experiments directly confirming this view ; he also thought that milk was simply chyle, not yet taken into the blood, that was deposited in the mammary glands.

for otherwise vinegar would digest bread as well as the stomach does. The next stage was to dispense with this distinction, on the assumption that the effervescent fermentation of wine was exactly the same kind of process as the effervescent solution of chalk in an acid. The professor of medicine at the now celebrated University of Leiden from 1658 to 1672, François Dubois (or de le Boë, and most commonly Sylvius, 1614–72) was the chief champion of the entire identity of physiological and laboratory chemistry. It only remained to assign the laboratory reactions that occur in the body ; Sylvius, adopting the facile chemical theory then popular that an acid-alkali neutralisation was the essence of all chemical phenomena, sought to find these antagonistic principles in the various secretions of the body.

In 1643 J. G. Wirsung at Padua had discovered the pancreatic duct and its secretion ; the English physician Thomas Wharton in 1655 and Steno in 1661 had described the submaxillary and parotid salivary glands respectively. Sylvius, observing that a large quantity of saliva is necessarily swallowed in eating, concluded that this was the first agent in digestion, reacting with the food in the stomach. The process was completed by a further, effervescent reaction between the bile and the pancreatic juice, since the former, Sylvius declared echoing van Helmont, " abounding as it does in a bitter and volatile salt ", could not but react violently with an acid. And that the pancreatic secretion was acid, seemed to follow from experiments by Sylvius' pupil de Graaf, who collected it by means of a tube inserted into the duct of a living dog. It was unfortunate that a mixture of the bile and the pancreatic juice did not show any effervescence *in vitro*, but it was thought their reaction might be more violent in the warmth of the viscera.

For nearly all the internal physiological functions which Borelli was to explain by purely mechanical actions, Sylvius had provided alternative chemical explanations. In respiration, for instance, " nitrous and subacid particles " of air served to condense the " rarified and boiling blood " leaving the heart ; while a kind of chemical precipitation enabled the kidneys to drain off the urine. Unfortunately, his chemical theory of digestion was undermined after only a few years by an experiment of the German physician Johan Konrad von Brunner. In 1682 he published an account of his observations on dogs from which he had removed the pancreas without seriously or immediately impairing their functions or endangering their lives. After all, it appeared that the pancreas was not essential to digestion.

Ingenious, promising, even on occasions exciting as the mechanistic approach to physiology was in the seventeenth century, it turned out—within the context of the age—to be no less sterile than the endeavour to apply its ideas in medical practice proved futile. Detailed attempts to contrive from fragments of mechanics and chemistry explanations of bodily function virtually came to an end in the first half of the eighteenth century ; Herman Boerhaave, the great medical teacher of Leiden who died in 1728, was the last partial representative of this tradition. It was supplanted by the conviction that the first and only starting-point for physiology is in anatomical exploration, avoiding all ideas derived from other sciences. The vivisection experiments in vogue during the second half of the seventeenth century were consequently only revived again in the nineteenth, under more stringent laboratory conditions, just as the primitive experiments on blood-transfusion of the 1660s were abandoned for nearly two hundred years. The ambition to mechanise biology as Galileo and Newton mechanised physics was made to seem hopeless by the failure of the crude mechanistic models of Descartes, Sylvius and Borelli. It was a sound ambition, linked as it was both with physiological experiment and with microscopic observation of structure, and one capable of realisation so far as the relatively simple mechanics of the skeleton were concerned. But physical science was as yet quite unable to yield the information required for understanding most of the more complex physiological processes. In the absence of a reliable theoretical architecture, to try to transport the still rudimentary notions of inorganic chemistry into the biological context was far-sighted, perhaps, but rash. The results could not withstand realistic examination.

Only in one aspect did mechanistic biology initiate, at this time, a line of attack that was to prove continuous. This was in the further study of respiration as a chemical phenomenon akin to combustion. Combustion and (as in metals) calcination were in their own right subjects no less obscure than other chemical problems ; but the experimental parallelism between breathing and burning was far clearer than others suggested by the physiological chemists, so that it seemed that whatever the explanation of combustion might turn out to be, the same would apply to respiration too. The observation that neither life nor a flame will persist in closed spaces was an old one, as was the useful discovery that if (in a mine, for instance) a candle-flame is extinguished, the air is unfit for breathing. Galen, and many others after him (including Descartes), had looked upon the heart as the seat of a vital flame, maintaining by means of the arterial blood the bodily warmth of animals. And Paracelsus, merely restating the classical distinction between life-giving atmospheric *pneuma* and the inert mass of the air, had affirmed that there is a small active fraction of the atmosphere essential to both breathing and burning. The first shadowy suggestion that this vital fraction has a nitrous character was made by a Polish alchemist, Michael Sendivogius, early in the seventeenth century.[24] There were apparently two reasons for making it : first, the only chemical effect, besides combustion, for which air was supposed to be essential was the formation of nitre from animal ordures ; and second, it was well understood that nitre was the active ingredient in effecting the peculiarly violent combustion of gunpowder.

Now the English physicians, intent on defending William Harvey's theory of the circulation of the blood, were particularly conscious of the problematic function of respiration. One of them, George Ent (1604–89), seizing upon Sendivogius' hint, declared boldly in 1641 that air was breathed chiefly for the sake of a nitrous quality in it that served to nourish the vital flame in the heart, as it did the ordinary flame. " The more nitre is contained in the air, the more strongly this flame burns."[25] Twenty years later the idea was a commonplace among the Fellows of the Royal Society, who now gave to it a corpuscularian dress, assuming that these particles of a nitrous kind were interspersed loosely among the more numerous inert ones of the true air, or were fixed among other particles in the solid nitre. The best-known statement is Hooke's, from *Micrographia* :

> the dissolution of sulphureous bodies is made by a substance inherent, and mixt with the air, that is like, if not the very same, with that which is fixt in Salt-peter, which by multitudes of Experiments that may be made with Saltpeter will, I think, most evidently be demonstrated.[26]

Before this, experiments plausibly evincing the necessity of air for life and combustion had been performed at Oxford in the late 1650s by Robert Boyle and reported in his *New Experiments Physico-Mechanicall, Touching the Spring of the Air, and its Effects* (1660). With the first of all laboratory vacuum-pumps he showed that a small animal and a flame expired equally quickly as the air was withdrawn from the receiver, although attempts to discover whether or not the animal would outlive a flame enclosed with it in air proved inconclusive. Nor did it prove possible to prove decisively that nitre *in vacuo* would support combustion, as gunpowder would burn under water. However, the absence of any change of pressure when a flame or an animal expired in a closed space was demonstrated, so that clearly the available air had not simply been used up. Nor did it seem that either was stifled by its own exhalations.

None of this experiment or speculation threw much light on the physiology of respiration, which was investigated in further experiments by the Royal Society. In one (formerly performed by Vesalius) Hooke in 1667 cut open the thorax of a dog so that the lungs collapsed ; the unfortunate animal was then kept alive by blowing air into them with a bellows. Upon Ent remarking that this showed only what respiration was not—" that the lungs did not

serve to promote by their agitation the motion of the blood "—
Hooke stated the positive (but mistaken) conclusion that the
respiration discharged " the fumes of the blood ".[27] Hooke's was
indeed an odd opinion for one who had addressed himself seriously
to the problem of combustion and studied the effect of access to air
on flames. At this point it was necessary to make more precise
experiments on the long-suffering animals, which Richard Lower
described in his *Tractatus de Corde* (" *Treatise on the Heart* ", 1669).
Lower went straight to the key question—the long known difference
in colour between venous and arterial blood. Suspecting that the
change took place in the lungs, he blocked the trachea of a dog and
found that blood drawn from the cervical artery was as dark as that
in the adjacent jugular vein. Borrowing Hooke's technique he
blew into the lungs with bellows while blood was forced through
them from the vena cava ; it issued from the pulmonary vein with
the bright, florid appearance of arterial blood. Moreover :

> That this red colour is entirely due to the penetration of
> particles of air into the blood is quite clear from the fact that,
> while the blood becomes red throughout its mass in the lungs
> (because the air diffuses in them through all the particles, and
> hence becomes thoroughly mixed with the blood), when
> venous blood is collected in a vessel its surface takes on this
> scarlet colour from exposure to the air.[28]

Indeed, Lower added, it was commonly known that venous
blood shaken up with air became completely red ; but how the
inspired air mingled with the blood in the lungs he could not explain.
(This part of his book would have benefited from acquaintance with
Malpighi's microscopic observations on the structure of the lungs.)
Here, then, at last was a physiological observation exactly duplicated
by a chemical action *in vitro* : respiration had the same effect as
mixing blood thoroughly with air. And Lower was in no doubt
that it was a " nitrous spirit " or " foodstuff " that passed into the
blood within the lungs, nor that when this spirit had again leaked
out from the circulating blood into the flesh of the body, the dark,

venous blood " that has lost its air " was left in the vessels. Pure air, therefore, was vital to health—" wherever a fire can burn, there can we sufficiently breathe "—but Lower did not go so far as to declare that the nitrous spirit was required for a physiological combustion and was used up in that process.

Within his limits Lower had solved the perennial problem of the purpose served by breathing, together with the more recent one posed by Harvey's discovery, of why the blood should be chased round and round the body at high speed. He had shown that a chemical change takes place in the lungs from which in some way the whole body benefits by the rapid transition of the blood ; he had also guessed what this change was, but had offered no suggestion to explain why an animal's survival depended upon its taking place.

The missing link in the chain was, of course, the retransformation in the tissues of the body of arterial into venous blood. What was the nature of the exhaustion or depletion that the blood leaving the heart suffered in its circulation ? This was a question that the seventeenth century, and indeed the eighteenth, left unsettled. One who offered some further hints was John Mayow (1645–79), a younger physician than Lower, also Oxford trained, who put together the various ideas and experiments current for fifteen years and more in two of his *Five Physico-Medical Treatises* (1674).[29] Mayow expressed his conceptions in the language of the mechanical philosophy with more imaginative dogmatism than any of his colleagues—for him many phenomena unrelated to either combustion or respiration were to be accounted for by the action of " nitro-aerial particles "—but he added no fundamentally new idea. The best thing he did was to make a really original experiment in which a mouse or candle was enclosed in a jar over water : Mayow observed that the water rose to fill about one-fourteenth part of the volume before the mouse died or the candle went out. In a neat way this proved that only a fraction of the air was used up in either case. However, he came to the inevitable, erroneous conclusion that the

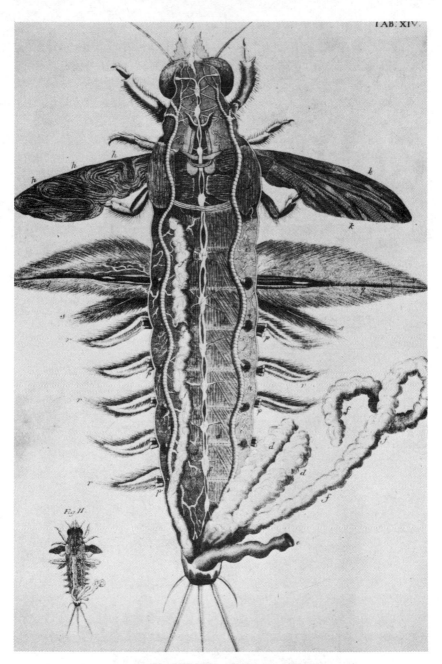

Swammerdam's anatomy of the May-fly from *The Book of Nature*.
The drawing represents an enlargement of about seven times

Experimental pneumatics: (*above*) von Guericke pumps water from a
sealed cask; (*below*) Boyle's second air-pump, ten years later, with
an animal placed in the receiver for an experiment in respiration

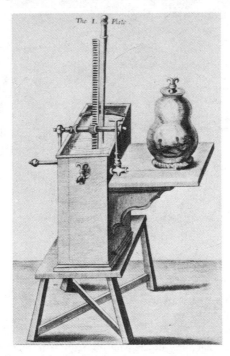

water rose because the elasticity of the enclosed air had diminished, for it was, he said, the presence of nitro-aerial particles that rendered the air strongly elastic. When these were used up the air lost its spring. Like Lower and others, Mayow thought it

probable that certain particles of a nitro-saline nature, and these very subtle, agile, and in the highest degree fermentative, are separated from the air by the action of the lungs and conveyed into the mass of the blood. For this aerial salt is so necessary to every form of life that not even plants can grow in soil to which air has no access.

Furthermore, it was no less likely that the

nitro-acrial spirit, mixed with the saline-sulphureous particles of the blood, excites in it the necessary fermentation* [which is the source of animal heat].³⁰

Finally, anticipating Borelli's theory, Mayow imagined that nitro-aerial particles issuing from the nerves and reacting violently with sulphureous ones in the substance of the muscles caused the muscle-fibres to contract. Hence (since this was a fermentation too), the heat produced by taking exercise and the saltiness of the sweat produced.

In Mayow's theory—which happens to sound far more plausible than he had any right to expect—a reason was proposed for the necessity that something in the air should constantly be absorbed into the body, even though the qualification of this something as nitro-aerial particles was highly speculative.³¹ Yet even Mayow refrained from appealing outright to the idea of physiological combustion, preferring the vaguer, inclusive term fermentation. Like others before (and many chemists of the day), he missed the importance of examining what comes out of a reaction, as well as what goes into it. Galen had thought that the expired breath carried sooty vapours from the vital flame in the heart, and Mayow in like fashion supposed that " the vapour of the blood agitated by fermentation is blown away by the lungs ", but no one tried to

* For Mayow combustion also was a kind of fermentation.

find out what actually was expired.[32] Gas chemistry was to be delayed for another eighty years.

Leaving aside the special studies of comparative anatomists, physiology signified in the seventeenth century human physiology ; even the entomologists were not completely clear that the processes of the organs of insects were quite different from those common to mammals. Indeed, the word physiology was rarely applied in the modern sense (Boyle's *physiologist* was, more broadly, a student of nature) ; its content was thought of as the theoretical part of medicine. Though experimental investigation was normally restricted to animals, the object was always to elucidate the comparable working of the human body. In the dramatic business of blood-transfusion, explored both in London and in Paris during the 1660s, after transfusions between living animals had been carried through with apparent success the experiments were actually extended to human subjects—until a victim died in Paris. In other cases attempts were made to support the conclusions drawn from experiments on animals by human dissection, especially dissection of pathological subjects, and clinical experience was often brought forward as evidence that natural processes occurred alike in man and his humbler companions. Practical medicine benefited neither from the experiments, nor the parallels so drawn ; in fact there was little significant advance in this most conservative art in spite of all the efforts of inquisitive medical men in the search for fresh understanding of health and disease. But clearly now, as implicitly for centuries past, man was regarded as no more than an animal in his physical structure. More deft, more sensitive yet blunt in his senses ; more physically adaptable than any animal yet primarily distinguished from the brutes (as the philosophers had always said) by his possession of a rational intellect, a fact beginning to be correlated specifically with his possession of a large and complicated brain.

" When I am describing the Brain of our Pygmie, you may justly suspect I am describing that of a Man," wrote Edward Tyson in his book on the anatomy of the chimpanzee (1699).[33] This was the first of the higher apes to be properly examined. Carefully comparing characters Tyson noted many that the chimpanzee shared with man alone, and that differentiated it from the monkeys : besides the brain, the vermiform appendix, the united bile and pancreatic ducts, the unlobed liver, the absence of cheek pouches. In other ways, however, the " pygmie " was closer to the monkeys than to man. Deciding that it was neither a hybrid nor a degenerate human form, Tyson wrote, " Our Pygmie is no Man, nor yet the common Ape ; but a sort of Animal between both ". Naturally he judged that his subject was " of the Ape kind " in spite of its structural conformities to the human type, but he found a creature that drew the likeness between man and brute sharper than ever before.

The distance between them was but measurable. In Addison's words :

Man is connected by his nature, and therefore by the design of the Author of all Nature, with the whole tribe of animals, and so closely with some of them, that the distance between his intellectual faculties and theirs, which constitutes as really, though not so sensibly as figure, the difference of species, appears, in many instances, small, and would probably appear still less, if we had the means of knowing their motives, as we have of observing their actions.[34]

If, in his anatomy of the chimpanzee, Tyson found no cause to support the ancient mythology of creatures half-men, half-animals, while at the same time recognising the dubiety of taxonomic frontiers, he did but echo the general feeling of his day. In their dealings with other races Europeans might practically deny their human status, but they never failed to recognise it theoretically. At this time they had had some two centuries of experience of primitive men, limited in their techniques, repugnant in their manner or life, and savage in their customs. Though enslaving them when opportunity occurred, and massacring them when they proved obstructive, Europeans seem nevertheless to have regarded primitive races as no less men than themselves, retarded rather by their environment and their attachment to sinful practices than by any inherent deficiency. They shared a touching faith that it was only necessary to Christianise and educate a savage to cause him to resemble in all essentials a simple countryman; all primitives possessed the power of articulate speech and no one doubted that even the rudest races had implanted in their souls an innate concept of God. The fiercest worshipped after their fashion, and all were doomed—if not converted in time—to eternal damnation with the fallen Christian. And that, after all, was the ultimate test for the possession of a rational soul, and of humanity.

No one had ever doubted that there were distinctions between men, though rather intellectual and moral than anthropological and anatomical. Nor was Tyson's near-human chimpanzee a shocking discovery. Man had his place in nature, not outside it. Living forms were universally believed to stretch in continuous procession from the most humble to the most sublime without gap or break, and it was as proper that there should be beings just inferior to man as others far superior. Among all the modes of existence from the inorganic to the angelic transitional forms were bound to occur, as the taxonomists admitted. Or as Locke wrote :

There are fishes that have wings and are not strangers to the airy region ; and there are some birds that are inhabitants of the water, whose blood is as cold as fishes . . . There are animals so near of kin both to birds and beasts that they are in the middle between both . . . There are some brutes that seem to have as much reason and knowledge as some that are called men ; and the animal and vegetable kingdoms are so nearly joined that if you will take the lowest of one and the highest of the other, there will scarce be perceived any great difference between them ; and so on until we come to the lowest and most unorganical parts of matter, we shall find everywhere that the several species are linked together, and differ but in almost insensible degrees.[35]

The taxonomists' bounds between groups were therefore quite artificial, or at any rate arbitrary. Man himself, occupying a middle position in the great chain of being—or as others thought a lowly one compared with the immense perfection of the angels—was a transitional form, on the frontier between bodies without souls and souls without bodies, uniquely possessing both. (In Kant's rough phrase, Newton appears to an angel as an ape appears to us.) Cartesian dualism made the issue all too clear : as a member of two orders of creation man was divided and distraught :

> Plac'd in this isthmus of a middle state,
> A being darkly wise and rudely great,
> With too much knowledge for the sceptic side,
> With too much weakness for the stoic pride,
> He hangs between ; in doubt to act or rest ;
> In doubt to deem himself a God or beast ;
> In doubt his Mind or Body to prefer ;
> Born but to die, and reas'ning but to err . . .[36]

Less prone than either philosophy or poetry thus to insinuate man into his logical place in a smooth ascending curve of existence, and more conscious of the morphological distinctions that the pretty theory of the great chain of being ignored, science was naturally drawn to treat man as the highest of the beasts, of which it knew something, not as the lowest of the angels of which it knew nothing. Mechanistic biology pressed the argument harder still. In the traditional theory of the great chain of being one scale of organic creatures terminated with man, and another scale of spiritual entities began with him. Similarly man alone embraced in himself—according to Aristotelian doctrine—the vegetative soul of plants, the sensitive soul of animals, and a rational soul shared with higher beings. The superiority of man had been in these terms categorised by philosophy, quite apart from its sublimation by religion. As against this, mechanistic biology, while making allowance for variations in complexity of organisation among living things, effectively denied that there was any distinction of principle between them, not excluding man himself. The argument from continuity that had formerly extended vital activity to the range far below man was now reversed ; if man, scientifically speaking, was an animal and if animals were complexes of elaborate physico-chemical mechanisms, then man was such a complex too. Life became a natural phenomenon, not totally dissimilar in fungus, animal or man ; even the mental faculties—emotion, expression, memory—because they were indubitably shared by animals in feeble measure, ceased to be in themselves indicative of man's superior status. Perhaps even reason failed as a final criterion separating man from beast ; if this were so, then man was left unique only by his possession of an immortal soul. In all other ways, if the logic of mechanistic biology were rigorously pursued, man was an animal and an animal a machine.

No one in the seventeenth century took his logic so far. The special character of man was unshaken, however illogically derived. Some defences of science still taught that the universe was made by God for man's enjoyment, so that it was incumbent upon him to explore his toy and tool, though this was an opinion now considered jejune by philosophers. But the argument that the creation was framed upon divine will and providence employed by Ray, Boyle, Newton and Bentley among many more, reflected belief in the special responsibility, or pre-eminence, of man hardly less directly. Indeed, it is at all times difficult to separate the scientific endeavour from implied belief in human pre-eminence, in one form or another. Impelled by the prevailing scientific impulse towards the mechanistic mode of explanation, biologists adopted a new view of the various objective likenesses and unlikenesses between men and animals. And if this did not at once affect man's idea of his place in nature, because his consciousness of superiority was firmly rooted in the philosophic concept of mind and the theological concept of soul, nevertheless, these furnished defences of pre-eminence, psychologically of the greatest importance, that were outside the framework of science. By the end of the seventeenth century scientific biology had quite renounced the notion of any special singularity in man, and had turned without regret to consider him as one species among others—though the taxonomists continued without hesitation to place him at the head of his class.

CHAPTER VIII

ELEMENTS AND PARTICLES

Had we senses acute enough to discern the minute particles of bodies, and the real constitution on which their sensible qualities depend, I doubt not but that they would produce quite different ideas in us, and that which is now the yellow colour of gold would then disappear, and instead of it we should see an admirable texture of parts of a certain size and figure.[1]

When John Locke in his *Essay concerning Human Understanding* distinguished between the primary qualities of bodies which are their inalienable attributes, and their secondary qualities which " in truth are nothing in the objects themselves, but powers to produce various sensations in us ", he did no more than follow the example of Galileo who had made just the same distinction in a famous passage of *The Assayer*. There he had postulated three primary qualities : " To excite in us tastes, odours, and sounds I believe that nothing is required in external bodies except shapes, numbers, and slow or rapid movements ". To these three Locke added two more, solidity and extension—which were indeed taken for granted in Galileo's *shape*. In this philosophy, then, if the actual mechanism of bodily sensation be left aside (Locke believed it to consist of the motions of the animal spirits stimulated by those of external objects), all the reality of the external world could be subsumed under one or other of these primary qualities ; and every kind of experience, save the internal experience of the sentient being, be reduced to combinations of their actions. The nature of the external world, the physical universe, could be perceived only through the effects of four variables : the shape, size, number and motions of bodies and their component parts.

This doctrine was far more novel in 1623 than it was in 1690, for when Locke wrote it had become a commonplace of scientific thought. Locke's use of it as the basis of his empiricist philosophy is the outstanding instance of the impact of seventeenth-century science upon philosophy. In Galileo's time the mechanical philosophy, as the explanation of phenomena in terms of matter and motion was called, was only just breaking away into independence from the ideas of Greek atomism, its most important root. Democritos and Epicuros had been known to the Middle Ages only through Aristotle's refutations of their ideas ; high interest had been aroused by the humanist rediscovery in 1417 of the Latin poem of Lucretius (" On the Nature of Things "), which for the first time gave a coherent and favourable account of the atomist's thought. The poem was widely read in the sixteenth century ; the association with atheism by which Epicureanism had been condemned was weakened ; the atom and the vacuum began to seem plausible scientific ideas.

The great intrinsic power of the atomic view of nature was that it reduced the explanation of change in things to the separation and combination of permanent entities, the atoms ; it was for the atomist to explain both how these separations and combinations occur, and how they produce the phenomena. In this the Greek atomists had not succeeded well. From its origin ancient atomism had failed to arrive at law-like statements or universal generalisations; rather when the atoms were not thought of statically they were imagined to have merely arbitrary movements. The great advance of the seventeenth-century mechanical philosophy was that it conceived of atoms (or more generally particles or corpuscles) as participants in a regular dynamic system, ultimately a mathematical system.

The idea that the particle has motion was not restricted to the movement that brings about rearrangement or redistribution, though this was one of the most important types of particulate motion. In framing explanatory hypotheses the attribution of a vibratory motion to particles was also useful, and so was the notion that particles are emitted in streams. These two last ideas added a whole new dimension to the atomist theory—to use anachronistic language one might perhaps say that the particle was now endowed with energy, and that it was now possible to seek explanations of events in the world in the various forms and levels of this energy. Already in 1623 Galileo—and he was not the first—had formulated the kinetic idea of heat :

> Those materials which produce heat in us and make us feel warmth, which are known by the general name of " fire ", would then be a multitude of minute particles having certain shapes and moving with certain velocities . . . Since the presence of fire-corpuscles alone does not suffice to excite heat, but their motion is needed also, it seems to me that one may very reasonably say that motion is the cause of heat.[2]

Galileo's was apparently an emissive theory : combustion causes fire-particles to be ejected, whose impact upon and penetration into the human skin causes the sensation of heat. This is a naive version. But the idea of emission could be applied, perhaps with better effect, to explain magnetic and electric attraction and even the radiation of light. At this stage, however, dynamical atomism, the core of the mechanical philosophy, was still embroiled with modes of thought derived from antiquity. The shape of atoms was thought to be significant in their effects, and the simple-minded notion that tough materials are made of atoms strongly hooked together, oily liquids of soft, round particles and so on persisted. The seventeenth century never quite succeeded in making the atom a completely neutral identity, varying only in size and motion, for there was always a tendency, in the endeavour to make atomist explanations as perfect as possible, to transfer to the ultimate particles of matter properties that belong only to the gross bodies composed of them.

For this reason the attempt to furnish in this new mechanical philosophy an alternative to the "forms" and "qualities" of Aristotelian physics ended in some cases only in redistributing the properties concerned. To the Aristotelian an acid was corrosive because it possessed the form of corrosiveness lacked by water and oil (just as opium possessed the *virtus dormitiva* ridiculed by Molière) ; to the seventeenth-century mechanistic chemical philosopher the acid owed its solvent power to its sharp, penetrating particles that got between and loosened those of the dissolved substance. Thus the abstract "corrosiveness" of Aristotle had simply been material-ised as "sharpness" in the particles of acid ; if the Aristotelian expression is an obvious tautology, the mechanical one was no more than a disguised tautology, as long as nothing was known inde-pendently about the shape of acid particles. And for chemical reasons an explanation of the dissolving power of liquids in terms of particulate shape rapidly proved inadequate. Hence many of the hypothetical explanations in terms of structure put forward by the mechanical philosophers seem at first no more informative than those that prevailed before ; an indetectable magnetic effluvium capable of passing through glass and copper is not necessarily a more plausible factor in magnetic phenomena than a magnetic attractive virtue, though it is certainly more picturable. The point is rather that the mechanical philosophy as a body of ideas offered a single pattern of explanation appropriate for both physics and chemistry, and enjoying strong internal coherence. It offered a structural theory, unified in principle at least.

The Aristotelian forms and qualities had no relation to physical structure, of course. They were immaterial, the characteristics variously impressed on basic matter which was in itself uniform, continuous and without characteristics. Just as the significance of a word depends on the shape of its letters, and not on the materials used to form them, so substances derived their properties from immaterial attributes, which in turn arose from the two simple antithetic pairs of qualities, hot–cold, dry–moist. In the development of this idea forms proliferated endlessly and without connection with each other, since every property demanded a new one. The total complex defined the body ; change the forms, and the nature of the body is changed. So by rusting iron loses the forms of rigidity, brightness and malleability and acquires new ones characteristic of a brown powder ; but the matter remains the same throughout the transformation. The bright iron is not just disguised as rust (though it can be recovered from it) because there has been a genuine substitution of one set of forms for another—not always, as in this case, reversible. The continuity of the matter involved is therefore of little significance, since it is to the change of form that attention must be directed. Moreover, though rusty iron is always brown and never green, the combination of forms in any such change is contingent rather than necessary ; in the Aristotelian theory each form is independent, so that (for example) the difference between lead and gold is not that between one material and another, but between one body that has the forms of yellowness, ductility, incorrosibility, etc., and another which has the forms of greyness, ductility, corrosibility, etc. Each form is a unit property.

The mechanical philosophers, on the other hand, seeking to re-interpret forms as modes of particulate structure and motion, were compelled to adopt a comprehensive point of view. The brightness, malleability, fusibility, magnetism and all other properties of iron, must be ascribed to a single particulate structure capable of producing all at once. What is more, simplicity demanded that whatever attributes of the particles of iron produced brightness should reappear in anything else that is bright like iron, and so on. A common pattern must necessarily run through the likenesses and differences between substances, from which the phenomena of nature arise.

As the first use of the kinetic aspect of the mechanical philosophy was in the explanation of heat, so the first use of its structural aspect was to explain colour, even though the fundamental concept of light that justified this use was still lacking. Bacon was the first to express the mechanical theory of apparent colours :

bodies entirely even in the particles which affect vision are transparent, bodies simply uneven are white, bodies uneven and in a compound yet regular texture are all colours except black; while bodies uneven and in a compound, irregular and confused texture are black.[3]

This was pure speculation—the empirical Bacon had made no experiments to support this theory—but in postulating four types of particulate structure corresponding to four different appearances under illumination Bacon anticipated the more positive theory of Newton. It was equally easy to imagine how corpuscular structure was reflected in the density, rigidity, fluidity, volatility and other properties of gross bodies. Experience offered ready models for the comparison of the microscopic texture of substances to that of plates, marbles or grains of sand, while useful analogies could be drawn with the regular shapes of crystals and of plant material seen under the microscope.

A third, more specialised application of the mechanical philosophy to chemical experiments leaned mainly on the idea of structure but made some use of the kinetic concept, as in the theory of solution. Earlier theories of chemical change had striven rather hopelessly to combine the belief that matter consists of three, four, or more elements with the infinite variation of Aristotelian qualities. Some at least of the seventeenth-century chemists were quite ready to abandon not merely the qualities but the elements as well. In the chemical version of the mechanical philosophy substances were composed of " corpuscles " in turn consisting of closely grouped particles. (There is an obvious but highly unreliable analogy with the later chemical concepts of molecule and atom.) The kind of corpuscles and particles, their shapes and motions, and their arrangement in structure all combined together to determine the gross properties of the substance—but exactly how could never be clearly defined. Chemical change could occur in two ways—either by rearrangement of the corpuscles without their disruption, or by rearrangement of the particles to form different corpuscles. Whether or not one or other of these changes took place depended upon the texture and motive power of the various reagents. Suppose, for example, two substances are each composed of pairs of corpuscles joined together (AB), (CD), then on heating a double redistribution might happen :

$$(AB) + (CD) = (AC) + (BD).$$

Though such a symbolic equation was never written the idea underlying it was certainly entertained ; it meant, in short, that a chemical process was capable of being represented by a mechanical model. But what caused such a change, so that the parts of the model fell into a fresh pattern ? Some materials dissolve in water, but not in acids, others vice versa ; some reactions require heat, others evolve heat, and so on ; here the kinetic aspect of the mechanical philosophy became relevant. The theory of chemical reaction, it seemed, must be one that explained how the component parts of the reagents were set in motion so that a redistribution could take place and new materials be formed. It was not easy to achieve this end.

After a half-century of rapid progress (c. 1620–70) the mechanical philosophy of nature was roughly stabilised until it was again reshaped by Newton's ideas, though these were never universally accepted. Its success owed much to Bacon and Galileo, and to lesser exponents of Lucretian atomism of their time, but its great authority derived from Descartes. Other than the human soul his science knew nothing but matter and motion. Whereas before 1644 appeal had been made to the model of material particles in motion in a few instances—heat, colour, magnetism—Descartes made it universally applicable to all events in nature, giving hundreds of examples in the *Principles of Philosophy* of the ways in which things might happen as a result of the interplay of particulate mechanisms. Moreover, except where Descartes failed to implement his plan, there was a thread of consistency running through them all. Contemporaries seized upon the hypothetical mechanisms of the

Cartesian world system as concrete illustrations of the mechanical philosophy, which they either imitated or criticised.

Descartes' pre-eminent influence in this respect over the physical theories of the later seventeenth century, almost exclusively concerned with the formulation of particulate mechanisms, has been needlessly obscured by two features of his system wherein he countered the tenets of ancient atomism. Descartes wrote always of particles, not atoms, because he believed in the infinite divisibility of matter. And he rejected the vacuum. Neither of these were points on which the mechanical philosophers who succeeded him felt it necessary to dogmatise, for the opinion of Boyle, that the distinction between particles and true atoms was metaphysical rather than physical was generally followed ; and the occurrence of true voids in nature could not be proved nor the presence of an aether (which promised great explanatory usefulness) disproved. After Descartes the mechanical philosophy did *not* signify (as it still had for Galileo) atoms-and-the-void ; it signified, more broadly, the use of kinetic, structural theories assuming that matter is composed of invisibly tiny particles. Newton, it is true, adopted a more strictly atomist position, but his views on the nature of matter were only published sixty years after Descartes' *Principles*. The only possible contemporary rival to Descartes was Pierre Gassendi, a voluminous exponent of Lucretian atomism. His philosophy was expressed in *Animadversiones in decimum librum Diogenis Laertii* ("Notes on the tenth book of Diogenes Laertius") of 1649 ; it lacked the combination of kinetic and structural elements typical of the developed mechanical philosophy.

The weaknesses of Cartesian physics from a dynamical point of view soon became apparent, and the manner in which Descartes exploited the concept of material particle was equally defective. He imagined unexplained changes from one species of particle to another, he accounted for the cohesion and rigidity of bodies composed of particles in an unsatisfactory fashion, and he made great play with idiosyncrasies of particulate shape. It was for his successors to follow the same plan of mechanistic explanation while improving on the execution of its details. In particular, Descartes in concentrating on physics had made no serious attempt to extend the mechanical philosophy to chemical phenomena. From about 1654 Robert Boyle (1626-91) applied himself to supplying this lack—not in a manner narrowly imitative of Descartes, however, for he interpreted the principles of the mechanical philosophy in his own broad and eclectic way. The *Physico-Mechanicall Experiments* of 1660, Boyle's first scientific publication, revealed very clearly his attachment to mechanical explanations though he declared his independence of both the atomist and the Cartesian schools. A manifesto announcing his ambition to bring chemistry within the scope of natural philosophy (that is, to render chemistry a branch of physics by explaining its phenomena on mechanical principles) was issued in *Certain Physiological Essays* ★ of the following year :

> I hoped I might at least do no unseasonable piece of service to the corpuscular philosophers, by illustrating some of their notions with sensible experiments, and manifesting, that the things by me treated of may be at least plausibly explicated without having recourse to inexplicable forms, real qualities, the four peripatetick elements, or so much as the three chymical principles.[4]

The endeavour was even more significant than it seems at first. For what Boyle was really attempting, as he perceived himself, was the creation of a single coherent theory of matter. He was almost as much a physicist as a chemist and in both thought and experiment he drew no sharp line between one branch of science and the other. He mingled evidence drawn from both in the same treatise, unity being implicit in the single problem : to explain how natural phenomena arise from the structure and motions of matter. Hitherto chemistry, rather than a science, had been a chaos : part laboratory technique, part craft skill, part pharmacology, part

---

★ The essay is entitled " Some Specimens of an Attempt to make Chymical Experiments useful to illustrate the Notions of the Corpuscular Philosophy ".

Detail from *The Alchemist*, by Antoni van Ostade (1610-1685), dated
1661. A "puffer" of the cruder sort is depicted

Seventeenth-century chemistry as seen in 1775: *The Discovery of Phosphorus* by Joseph Wright

Aristotelian or mystical philosophy. The pretence that the empirical facts discovered by the practical chemists were comprehended adequately under the explanatory principles of Aristotelian physics had virtually collapsed in the Middle Ages, when the specialised alchemical notions of qualitative transformation came into vogue. Since then, despite the reluctance of all rational men acquainted with chemical operations to commit themselves to the alchemists' fancies, which led only to absurdity, double-talk and deceit, there had been no worthwhile alternative. The only resort was to take a limited and pragmatic view of the usefulness of the " spagyric principles " (or Paracelsan chemical elements, salt, sulphur and mercury) while concentrating on what was solid, the purely practical manipulations of the chemical art.

Traditional philosophy had ignored chemistry altogether, an attitude reinforced by the post-Galilean emphasis on mathematics in physical theory. Descartes, indeed, was generous in finding any room for chemical change in the system of nature. About the middle of the seventeenth century, while the physicist was described as a natural philosopher and regarded primarily as a man of ideas, verified by a careful experimental demonstration here and there, the chemist was typically looked upon as a man rich in experience drawn from the laboratory but naive in his manner of accounting for his results. It was no accident that Boyle, a major pioneer of experimental physics, commenced his interest in science with chemistry. The language in which typical chemists attempted to explain why things happen as they do was peculiar to their class and barely intelligible to the uninitiated, who found the concepts to which the chemists appealed equally bizarre. Not a few of the practical ones, like Nicolas Lemery (1645–1715) in his *Cours de Chimie* (1675), preferred to restrict theorisation to a preface and fill the body of their book with descriptions of processes, their products and their applications to medicine. Others, however, such as J. R. Glauber (1604–70), did look for the reasons underlying chemical experiments in the redistribution of elementary principles—" salt ", the principle of dryness and fixity ; "sulphur", the principle of volatility and combustion ; and "mercury", the principle of fluidity and metallicity. Their theory turned on a materialisation of the Aristotelian qualities so intricate that none but specialists understood it.

As Glauber emphasised, these "principles" were not the common substances of the same name :

> most false is the foundation of such as imagine that Metals have their Original from common running [mercury] and burning sulphur (each being a semi-metal) tis indeed certain that metals are born of [mercury] & sulphur but not the common, but such [as] aforementioned, viz. Astral, a sulphureous, warm, dry and spiritual soul, and terrestrial viscous water, from whose mutual conjunction (as of Male and Female Seed) all Metals are born. That Erroneous opinion hath been the cause of many labours on [mercury], and they are not a few who have wasted all they had by this, their Philosophy.[5]

And though the same principles were constituents of all matter, this was a truth that could hardly be made to appear to the untutored eye :

> . . . many not believing in Nature and Art, will only say that it could never be that Wood should be transmuted into Gold, and Stones into Gemms. And although the latter should be in some sort credible, by reason of the alliance or likeness between Stones and Gemms, nevertheless the former, by reason of the great disagreement between Wood and Metals, inasmuch as they belong to different kingdoms, they object, exceeds all Belief . . . But such things are wont to move no admiration in Philosophers, much less any doubt, seeing that they are not ignorant how great familiarity there is of the Vegetable Kingdom with the Mineral, both which have their rise from the same Subjects, viz. Salt and Fire,* which their Anatomy doth

---

* Note the contradiction with what Glauber said before. Later he even speaks of metals arising from a *sulphureous salt* !

clearly demonstrate. And although this may exceed the capacity of some, yet it doth not thence follow that it is not true. If all things were to be spoken truly and openly, I confess there would be found a very small number of those who rightly understand Nature . . . Therefore it is no wonder that the Secrets of Nature should be hid, and by the ignorant and unskilful of this sort be taken for Fables and foolish Whimsies, which their Sheeps Brains cannot reach.[6]

Glauber wrote most of his books between 1648 and 1661 ; he was not—for a chemist—a very obscure writer, but he reflected well enough the tendency of the more esoteric chemists to shroud their ideas in a fog of allusive verbiage and contradictory technicalities. As Boyle was to point out, the peculiar language that Glauber defended with the age-old plea that the complexities of nature can only be expressed in complex forms of words, really had no definite meaning at all. It was not so much an attempt to express ideas as a barrier to the development of ideas. Van Helmont is the supreme example of this, for even contemporaries scratched their heads over his bulky folio, and no one has really been able to penetrate his meaning since. Concepts loosely and grandiosely conceived, freely adapted to every circumstance, cease to be expressive of anything but emotion. Despite all this, empirical chemistry had made great strides, largely as an apprentice to medicine—Glauber's own works were important vehicles in this respect. In Boyle's youth it was far easier than ever before to discover what the reagents of chemistry are, how to prepare and manipulate them in various ways to bring about a desired reaction, and to discover what the products of this reaction were supposed to be. Ideas on this subject were, of course, often as misleading as the chemical terminology ; just as butter of antimony and sugar of lead had no relation to the table, so neither the " burning spirit of Saturn [lead] " nor the " mercury of life " contained any metal. Such mistakes were inevitable, but at least the sound, practical body of laboratory experience had increased enormously since the time of Libavius

(1540–1616). Chemical experience was ready to be organised—or so it seemed.

Boyle's first attempt to bring organisation into chemistry has been curiously misunderstood, and probably overestimated. The composition of the *Sceptical Chymist* (1661) was begun some years before and the book has all the faults of Boyle's youthful endlessly wordy, shapeless style cast into the form of an unlikely dialogue. He made the mistake of adopting a negative pose, as though intending no more than to point out the insufficiencies of the philosophers' four elements and the chemists' three principles without making way for an alternative opinion. Boyle's account of elementary principles and of their role in chemical theory contained nothing new nor, though he gave some good knocks against the evidence presumed to prove their existence in all substances, did he succeed in effacing belief in their reality. Contrary to popular mythology Boyle drove neither the Aristotelian elements nor the spagyric principles out of chemistry. What, more accurately, he was about was the undermining of the Aristotelian doctrine of forms and qualities and of the chemical explanations that depended upon them. In so doing he appealed to the indestructibility of matter, from the beginning " actually divided into little particles of several sizes and shapes variously moved ", and then " associated into minute masses or clusters . . . not easily dissipable into such particles as composed them ".[7] He drew attention to the persistence of the identity of gold and quicksilver, which was only lightly concealed through countless chemical changes, as evidence of the inviolability of the " clusters " of the metals, that could be recovered unchanged at the end. Clearly the varying qualities of the various compounds had in no way altered the substance of the metals. But the allusions to the mechanical philosophy in the *Sceptical Chemist* are not elaborately worked out and it requires careful reading to discover that Boyle's rooted objection to the whole concept of element arose from the impossibility of finding out how many kinds of fundamental particle there are; for to his mind (very logically) only these could be considered true elements. All else, every accessible substance pure and impure, was in his view made of " mixt " corpuscles composed of several particles ; hence not necessarily permanent, and certainly not elementary. The mechanical philosophy had made the detection of chemical elements impossible and their very existence unlikely.

How Boyle spent his life applying his powerfully creative if unsubtle mind to the extension of the mechanical philosophy is plain even from the titles of some later books in which (unlike the *Sceptical Chymist*) his own experimental researches were plainly described. With an obvious insistence they show that Boyle was far from being a sceptic, and that he had definitely embraced the corpuscular theory of matter :

> *The Origins of Forms and Qualities according to the Corpuscular Philosophy* (1666) ;
>
> *Some Thoughts about the Excellency and Grounds of the Mechanical Hypothesis* (1674) ;
>
> *Experiments, Notes &c about the Mechanical Origin or Production of divers particular Qualities : Among which is inserted a Discourse of the Imperfection of the Chymists' Doctrine of Qualities* (1675) ;
>
> *An Essay of the Great Effects of even Languid and unheeded Motion* (1685) ;
>
> *Of the Reconcileableness of specific Medicines to the corpuscular Philosophy* (1685).

Elsewhere, as in the *Experimental History of Colours* (1664), qualities which Boyle disposed of by assigning them to their mechanical origin, he made free use of his notion of corpuscular architecture :

> I am apt to suspect [he wrote] that if we were sharp-sighted enough, or had such perfect Microscopes, as I fear are more to be wish'd than hop'd for, our promoted Sense might discern in the Physical Surface of Bodies both a great many latent Ruggidnesses and the particular Sizes, Shapes, and Situations of the extremely little Bodies that cause them, and perhaps might perceive among other Varieties that we now can but imagine,

how those little Protuberances and Cavities do Interrupt and Dilate the Light, by mingling with it a multitude of little and singly undiscernable Shades . . . according to the Nature and Degree of the particular Colour we attribute to the physical Object. . . .[8]

In this case Boyle's inability to say exactly how the texture of the reflecting surface modified the light received by the eye to render it coloured was repaired by Newton, who discovered that a surface reflects more of some of the constituents of white light than others. In the *Origin of Forms and Qualities* Boyle presented his most complete theory of the particulate structure of matter and its effect upon phenomena. Matter and motion he described as " the two grand and most catholick principles of bodies ", the species of motion being in part determinant of the type of the body. These several species of motion Boyle regarded as imparted to the particles of matter by God in the beginning, and though he held that the particles were theoretically divisible—by the omnipotent power of God, for instance—in practice these *prima naturalia* always remained undivided. As for some particular qualities of material bodies, Boyle taught that heat resulted from an agitation of the particles, the elasticity of true air from the springiness of its particles, magnetism from the emission of a magnetic effluvium, and so on. On occasion Boyle worked out the explanation of the quality in some detail, in the manner of Descartes ; in the case of heat, for instance, Boyle observed that a nail only grows hot under the hammer when driven up to the head, for then the whole effect of the hammer blows is spent on agitating the particles of iron.

To chemistry Boyle constantly returned after 1661 for illustration of the applicability of the mechanical philosophy to experimental science. For all his hundreds of experiments with the air-pump chemistry was the science he loved best, and of which he had the richest and most varied experience. And though he brought a physicist's outlook to chemical problems, believing that knowledge of the structure of substances held the means to resolve them,

Boyle was no cabinet theorist. His technical skill and preceptiveness were quite equal to his theoretical capacity, which was unexcelled in his day. Shrewdly, he rejected nearly all the prevailing chemical fashions, not just the three principles, and recognised (correctly) the possible faults of certain of his own procedures. Unlike most chemists of the day Boyle knew the importance of obtaining pure reagents, and the importance of measurement in chemical experiments, though what he could do in that respect was limited by his techniques. He devised original methods, such as the use of colour-indicators to test for acidity or alkalinity. On some occasions he made " control experiments " to check the main trial. From long experience he acquired both a sound knowledge of analysis, in so far as this could then be attempted, and a methodical procedure for investigating the properties of new substances, such as phosphorus (which Boyle found out how to prepare from the merest hint that the starting material was urine). Thus, when Boyle tried to explain what happened in a chemical reaction he had in many cases a roughly correct idea of the relation between the end-products and the initial materials—except, of course, where gases were concerned.

Broadly speaking Boyle wrote of chemical substances in two ways. He imagined composite bodies—and all reagents were composites in his opinion—to consist of two or more " parts " variously characterised (from their properties, or their origins) as saline, sulphureous, acid, metalline, earthy, alkaline, aerial and so on. Here Boyle wrote in a manner not very dissimilar from that of earlier chemists. Where he went beyond them was in identifying such " parts " of substances with features of their corpuscular structure. Thus when silver was dissolved in spirit of salt (hydrochloric acid) *Luna cornea* (silver chloride) was formed, he thought, by the penetration of saline particles (deriving from the salt with which the acid was made) among the particles of the silver, so that the new material *Luna cornea* must be composed of two kinds of particle at least. Similarly Boyle held that sal ammoniac (NH_4Cl) consisted of particles drawn from common salt ($NaCl$) united to others drawn from spirit of urine (ammonia, NH_3)—correctly enough, as it happens. Such a composition of sal ammoniac, suggested by the manner of its preparation, further explained the reappearance of the "saline spirit" (ammonia) when the sal ammoniac was heated with an alkali ; since in that event, by another separation and rearrangement :

the Volatile Urinous Salt . . . that abounds in the Sal Armoniack . . . is set at liberty from the Sea-salt wherewith it was formerly associated and clogged, by the Operation of the Alcaly, that divides the Ingredients of Sal Armoniack, and retains that Sea Salt with it self.* 9

A chemical reaction was for Boyle a realignment of the component particles or corpuscles (he was not consistent in making a distinction between the two) of the reagents, which he came near to attributing to variations of affinity between them. (Apart from this hint he had little to say about the forces that cause the realignment to take place.) As a result particles that were released without finding a fresh association fell down as a precipitate, or rose as a sublimate or vapour, if volatile. But Boyle did not believe that all the redistributions of particles effected by nature or the chemist were reversible ; for instance, it was impossible to recover from sugar of lead the acid spirit of vinegar that (he rightly thought) had combined with lead to form it. Nor could the factitious product glass be analysed into its components. In some processes such as these the ultimate particles of matter had been reforged into new corpuscles, so much more strongly bound together than the old ones that had been destroyed, that they could not in turn be disrupted. Some corpuscles, like those of gold, seemed to be unbreakable, for the metalline properties could not be permanently removed ; nevertheless, Boyle's theory of matter taught him that in principle—though probably not in practice—all substances could be broken down to the particulate level and reconstituted. Even the transformation of

* I.e. sal ammoniac   ammonia

$$(v.u.s + s.s) + A \rightarrow v.u.s + (s.s + A).$$

gold was not out of the question, therefore. After reporting a misleading experiment in which water was apparently turned into earth, Boyle wrote :

if the [Aristotelian] elements themselves are transmutable, and those simple and primitive bodies which nature is presumed to have intended [for] stable and permanent ingredients of the bodies she compounds, may by art be destroyed and reproduced; why may not the changes which happen in other bodies, proceed from the local motion of the minute or insensible parts of matter, and the changes of texture consequent thereto ? [10]

Boyle was also keenly interested in other results of the " changes of texture " produced by chemistry, notably alterations in the qualities or physical properties of bodies. Lead, he pointed out, might be either a greyish metal or a brilliant red powder ; mercury a dense liquid or a white corrosive solid. He was particularly delighted with one experiment showing

that we need not suppose, that all Colours must necessarily be Inherent Qualities, flowing from the Substantial Forms of the Bodies they are said to belong to, since by a bare Mechanical change of Texture in the Minute Parts of Bodies, two Colours may in a moment be Generated quite *de novo*, and utterly Destroy'd. [11]

He simply dissolved mercury sublimate in water and added oil of tartar, whereupon the solution turned a deep orange colour from the precipitate formed (mercuric oxide) ; with a few further drops of oil of vitriol to neutralise the solution it became colourless again. (Boyle knew that spirit of urine used instead of the oil of tartar gave a white precipitate, so that the experiment distinguished animal from vegetable alkali.) This was no chance observation either, for Boyle had reasoned that since a solution of mercury in nitric acid gave a yellow precipitate, one of the sublimate would do the same, because both contained saline particles. The colours displayed by corpuscles hidden in a chemical solution, therefore, were as much dependent on their mechanical constitution as the colours of solid bodies. And that, conversely, colour-changes were indicative of mechanical constitution was evident both from this experiment and those Boyle made on acid-alkali colour indicators.

Hence in Boyle's theory, *both* the chemical and the qualitative properties of bodies derived from their mechanical texture ; if that— the underlying physical reality—were modified, then all the former properties would be changed too. Experiments on · coloration demonstrated to him decisively that the facts of both physics and chemistry were accounted for by the corpuscular philosophy. Boyle's mechanical hypotheses of qualities and his mechanical hypotheses of chemical reaction were two sides of the same coin.

Perception of a profound truth may impose its own limitations on vision. Both before and after Boyle elements have been defined as substances that cannot be resolved into component materials. Chemical transformation of one element into another becomes by such a definition impossible, since no two elements can have any- thing chemically common to both. In Boyle's essentially physical theory of matter, however, as in the modern theory, chemical substances were held to be all formed from the same ultimate particles of matter, which for Boyle were at least sometimes subject to chemical operation and rearrangement. Thus, the possibility of a genuine chemical transformation of matter existed as a natural consequence of the mechanical philosophy as both Boyle and Newton expounded it ; that in this respect they were (in the judgement of posterity) half-right and half-wrong is the price they paid for trying to make chemistry a branch of physics. This endeavour, whose metaphysical justification was the assumption that all matter is ultimately one, diminished their attention to the important fact that chemical identity does persist through chemical change. It was a fact that both Boyle and Newton recognised in some instances, but they failed to perceive its theoretical force, and so to arrive at Lavoisier's use of the element concept. Nor, and this was immediately crucial for their theory, could they progress further to the idea that a chemical element is a substance composed of identical particles of a kind peculiar to that substance, exploited so fruitfully by Dalton a century and a half later. The early nineteenth-century chemical atomism demanded the postulate (which physics was later to falsify) that one substance is composed of only one kind of particle—the simplest possible law ; but the form in which the mechanical philosophy was adapted to chemical phenomena in the seventeenth century made the apprehension of that simple law impossible. For oddly enough it would have destroyed the principle of simplicity of physical structure which it was the main object of that philosophy to assert.

Boyle's chemical philosophy was not widely imitated outside England, though his authority as a practical writer on chemistry was extremely high. Chemists less philosophical than he preferred to speak of the " parts " of substances, and these they sought to identify with three, four or more elementary principles. Like their predecessors they aimed at explaining chemical matters in a purely chemical, qualitative language innocent of ideas borrowed from physics ; this was especially true of the German chemists among whom, headed by J. J. Becher (1635–82) and G. E. Stahl (1660–1734), the prevailing chemical theories of the early eighteenth century developed. The trend of ideas that produced the phlogiston theory was hardly touched by the rationalist, mechanistic philosophy that had influenced Boyle, and the whole tradition of physics. But the phlogiston theory was not dominant until after 1730, and it did not renounce Boyle, nor the particulate concept of matter. Rather it regarded the latter as irrelevant to the explanation of chemical phenomena.

Chemical theory after Boyle's time was much concerned with explaining combustion : indeed, Stahl's phlogiston concept is often portrayed as though it were wholly concerned with burning—which it was not. Boyle himself had commonly taken his examples from wet-chemistry and he had furnished no coherent mechanical hypothesis of fire and flame. There were, however, several English scientists of his generation who hopefully directed the mechanical philosophy to the explanation of combustion, though they were less interested in the qualitative changes produced by it than was Boyle. As chemists had done before they identified the volatile principle occurring in all combustibles as " sulphur " ; the novelty of their view lay in the fact that they regarded burning as an inter-action, between these sulphureous particles and certain particles present in the air and in " nitrous " bodies (cf. p. 205). The site and manifestation of this interaction was typically a flame, where the parts of the combustible body were dissolved by the air. In Hooke's words (1665) :

the Air which encompasses very many and cherishes most bodies is the menstruum, or universal dissolvent of all sulphureous bodies . . . this action it performs not, till the body be first sufficiently heated, as we find requisite also the dissolution of many other bodies by several other menstruums [solvents] . . . this action of dissolution generates a very great heat [and] is performed with so great a violence that it produces in the diaphanous medium of the Air, the action or pulse of light.[12]

Mayow's view was much the same, except that he distinguished the nitro-aerial particles from the others composing air as being those active in combustion :

Fire [he wrote] is nothing else than an exceedingly vigorous fermentation of nitro-aerial particles and sulphureous particles in mutual agitation.

Hence, these theorists supposed that a burning substance was divided into sulphureous particles that entered the flame, and others, inactive, that went into the ashy residue. Fire, as always before Lavoisier, was an analysing agent ; change the term sulphur to phlogiston (which was no more than a new name for the old combustible principle) and the account is not so very different from Stahl's. The fate of the particles that entered into combustion in the flame was never clearly defined ; presumably the sulphureous ones ultimately formed soot or smoke or vapour, while according to Mayow the nitro-aerial particles simply disappeared, or emerged as heat and light. This accounted for Mayow's observation that the volume of air was reduced when combustion took place over water,* but with the awkward consequence that the principle of the conservation of matter accepted by other seventeenth-century scientists was tacitly abandoned.[13] Boyle, of course, had always held that however much the configuration and arrangement of particles was altered in a chemical reaction, the quantity of matter remained the same ; though for him " matter " included fire-particles that rendered the calces (oxides) of metal heavier than the virgin metal.

Newton was the last to theorise in Boyle's manner on the structure of matter, his remarks so transforming the mechanical philosophy that the eighteenth century drew upon him, rather than from Boyle. As a result the extent to which Newton had followed the pattern created by his older contemporary went unappreciated, although Boyle's influence on Newton in this and other respects was certainly very considerable. Newton took copious notes from Boyle's writings, from which he derived an important part of his scientific education, and it was probably reading the *Experimental History of Colours* that set Newton off on his own optical experiments. His theory of matter, developed over many years, was expressed in unpublished drafts of the 1670s, in letters written to the Royal Society and to Boyle himself, in passages written for the *Principia* that he left unprinted and in less explicit allusions in that book, and in the *Queries* appended to *Opticks* composed during the first decade of the eighteenth century. Newton's final word on this topic was in the last great scientific statement he ever made, the General Scholium added to the *Principia* in 1713.

Through all these documents runs a common thread of ideas. Newton was a more literal atomist than Boyle :

it seems probable to me [he wrote] that God in the Beginning form'd Matter in solid, massy, hard, impenetrable, movable Particles, of such Sizes and Figures, and with such other

* Since he was unaware that the carbon dioxide formed dissolved in water.

Properties and in such Proportions to Space, as most conduced
to the End for which he form'd them. . . .
These were unchanging. And so, that matter might not wear away
or alter,

the Changes of corporeal Things are to be placed only in the
various Separations and new Associations and Motions of
these permanent Particles.[14]

Like Boyle, Newton thought that these original particles were
compounded into other corpuscles and bodies, some less easily
disrupted than others. Many times he insisted that particles were
not thrown together haphazard like a heap of stones ; they must be
arranged in regular patterns (as crystals proved), perhaps into lines
first, then networks and three-dimensional figures.* From their
particulate configuration some of the properties of gross bodies,
such as transparency and opacity and chemical solubility, arose ;
Newton observed, however, that all bodies (even the densest, gold)
are translucent in thin films or when divided by solution. From
this, and other arguments, he concluded that the particles are never
packed close together ; in all substances, therefore, the proportion
of solid matter must be small compared with the volume of empty
space.

By implication, at least, Newton believed that all the fundamental particles are alike—though not, of course, the higher-order
corpuscles.

For the matter of all things is one and the same, which is
transmuted into countless forms by the operations of nature,
and more subtle and rare bodies are by fermentation and the
processes of growth commonly made thicker and more
condensed.[15]

(Here Newton, like Boyle and other earlier philosophers, drew an
analogy from the formation of solid bodies out of liquids : crystals

---

* Kepler published the first figures showing how spherical particles might be
arranged to give regular geometric figures, like those of snowflakes (*Strena*, 1611).
The notion was taken further by Hooke (1665) and Huygens (1690).

from solutions, plants from water.) Since all matter is ultimately the same, and because many surprising qualitative transformations are known, almost any transmutation of properties was imaginable through the endless and subtle modifications of corpuscular structure :

> The particles of such bodies can very easily be agitated by a vibrating motion and this agitation may last a long time (as the nature of heat requires). Through such agitation, if it is slow and continuous, they little by little alter their arrangement, and by the force causing their coherence they are more strongly united, as happens in fermentations and the growth of plants, by which the rarer substance of water is gradually converted into the denser substance of animals, vegetables, seeds and stones. If however that agitation is vehement enough, they will glow through the abundance of the light emitted, and become fire.[16]

All this is very close to Boyle's mechanical philosophy, except that Newton emphasised the vibrating motions. Where the latter went far beyond Boyle was in discussing the *forces* between particles and—in one instance—outlining the whole physical system that one such force produces. In the *Principia* Newton displayed the mechanical architecture or "system of this visible world, so far as concerns the greater motions that can easily be detected", especially the regular orbital motions of the heavenly bodies.

> There are, however [he added in an unpublished passage], innumerable other local motions which on account of the minuteness of the moving particles cannot be detected [directly], such as the motions of the particles in hot bodies, in fermenting ones, in putrescent and growing bodies, in the organs of sensation and so forth. If any one shall have the good fortune to detect all these, I might almost say he will have laid bare the whole nature of bodies so far as the mechanical causes of things are concerned. I have least of all undertaken the improvement of this part of philosophy.[17]

In spite of the modest disclaimer, Newton did work out his ideas on this part of philosophy in considerable detail, and even printed a fairly full account of them in Query 31 of *Opticks*. There he declared his belief that material particles " have not only a *Vis Inertiae* . . . but are moved by certain active Principles, such as is that of Gravity, and that which causes Fermentation, and the Cohesion of Bodies ". For, as he explained elsewhere, Nature is everywhere uniform :

> whatever reasoning holds for greater motions, should hold for lesser ones as well. I suspect that the former depend upon the greater attractive forces of larger bodies, and the latter upon the lesser forces, as yet unobserved, of insensible particles. For, from the forces of gravity, of magnetism, and of electricity it is manifest that there are various kinds of natural force, and so one cannot deny that there may be many of them. It is very well known that great bodies act upon one another mutually by those forces, and I do not clearly see why lesser ones should not act upon one another by similar forces.[17]

The path of Newton's thought is not hard to follow here. Experiments with magnets and electrified bodies showed that they were able to move (and must therefore exert a force upon) other bodies at a distance from them. Further, his own study of gravity in the *Principia* had demonstrated that all matter exerts an attractive force on all other matter, also at a distance. (These were empirical facts, whatever the nature of " attraction " or " force ".) Since these forces could scarcely reside in large masses of matter unless their constituent particles also possessed them, Newton believed that the forces of nature did belong to the particles of matter ; in fact some of the mathematical reasoning of the *Principia* itself rested on the postulate that the gravitational force (whatever it is) extended down to the fundamental particles. But the mechanical philosophy also assumed, and Newton accepted the assumption, that other phenomena, such as those of heat, chemical change and so forth, resulted from the motions of particles, even though these microscopic motions caused no visible displacement.* Newton reasoned that these motions also were brought about by forces associated with the particles of matter. Thus any given particle might be endowed with several different forces by which it acted upon other similar particles : the static force of cohesion, by which bodies resist fracture ; the dynamic forces of gravity, magnetism and electricity ; the force of chemical change ; and the kinetic forces of heat and light. Or possibly these were all various aspects of a single force.

Newton considered attentively the chemical tensions that re-arrange particles when (for instance) a metal is dissolved in an acid. Though undetectable directly, he regarded the chemical force as very powerful on the minute scale of corpuscular archi-tecture. He suggested, moreover, that many chemical reactions could be accounted for by supposing that this force varied from one substance to another ; so, he wrote, when zinc precipitates iron from a solution of ferric nitrate, or iron similarly precipi-tates copper from cupric nitrate :

> or a Solution of Silver dissolves Copper and lets go the Silver, or a Solution of Mercury in Aqua Fortis [nitric acid] being poured upon Iron, Copper, Tin or Lead, dissolves the Metal and lets go the Mercury ; does not this argue that the acid Particles of the Aqua Fortis are attracted more strongly by the [zinc] than by Iron, and more strongly by Iron than by Copper, and more strongly by Copper than by Silver, and more strongly by Iron, Copper, Tin and Lead, than by Mercury ? [18]

Thus Newton added to Boyle's kinetic theory of chemical change the idea of chemical attraction, or affinity, as the force that effected the necessary rearrangement by motion of the constituent particles. Other experiments seemed to demonstrate clearly the force of cohesion. When two solid plates, ground flat, are pressed together a considerable force is required to separate them again ; when a

* Except in the case of heat, thermal expansion and contraction being detectable as movement.

drop of oil is placed between two sloping strips of glass, inclined at an angle, it will run uphill between them into the angle. These effects, Newton reasoned, were occasioned by the same force that causes the particles of solid to resist separation in fracture.

More generally, perhaps taking as his model this time the polarity of magnetic and electric forces,* Newton believed that there were two types of force in nature, the attractive and the repulsive :

> one of these impels adjacent particles towards one another, this is the stronger but decreases more quickly with distance from the particle. The other force is weaker but decreases more slowly and so at greater distances exceeds the former force ; this drives the particles away from each other.[19]

For it is hard to press two solids into close contact, and dry powders cannot be made to cohere by the pressures that Newton could command. But when contact had been made—as by wetting a powder and then drying it hard—the cohesive force came into play. Moreover, Newton thought that there might be an actual repulsion involved in chemical reactions, besides a more or less strong attraction. He was not always consistent in the details of his theory of matter as he explained it on different occasions, so it cannot be too closely examined ; he did not say how many such natural forces between particles there are, and may perhaps have believed that in the last resort they were all manifestations of one single force. Such, at any rate, is the sense of the mysterious final sentence concluding the General Scholium to the second edition of the *Principia*, as explained by some of Newton's manuscript notes :

> Something might now be added about a certain very subtle spirit that pervades all dense bodies and is concealed in them, by whose force and actions the particles of bodies attract each other when separated by very small intervals or cohere when contiguous ; and by which electric bodies act at greater distances,

---

* Newton knew of electric attraction and repulsion, not the polarity of positive and negative electricity.

both repelling and attracting neighbouring corpuscles ; and by
which light is emitted, reflected, refracted and inflected, and
heats bodies ; and by which all sensation is stimulated, and the
limbs of animals are moved at will—for this is done by the
vibrations of this spirit transmitted through the solid capilla-
ments of the nerves from the external organs of sensation to the
brain, and from the brain to the muscles. But these things
cannot be explained in a few words, nor have we at hand a
sufficient number of experiments by which to determine and
demonstrate the laws of action of this spirit accurately, as ought
to be done.[20]

For the spirit to which Newton referred, with that strange depth
of insight so often encountered in his work, was electricity. In
other words, he suggested that the universe and even the living
things in it were composed of hard, inert particles of matter and
electrical energy. Nothing but these, the souls of men, and the
omnipotent power of God.

CHAPTER IX

EXPERIMENTAL PHYSICS

As in Mathematicks, so in Natural Philosophy, the Investigation of difficult things by the Method of Analysis, ought ever to precede the Method of Composition. This Analysis consists in making Experiments and Observations, and in drawing general Conclusions from them by Induction, and of admitting of no Objections against the Conclusions, but such as are taken from Experiments, or other certain Truths. For Hypotheses are not to be regarded in Experimental Philosophy. And although the arguing from Experiments and Observations by Induction be no Demonstration of general Conclusions ; yet it is the best way of arguing which the Nature of Things admits of, and may be looked upon as so much the stronger, by how much the Induction is more general.[1]

The " Newtonian synthesis " was not confined to the marriage of dynamics and astronomy, grand though that was ; potentially Newton's synthesis was the comprehension of all physical science within a single theory of matter, at once mathematical and mechanical, from which all the phenomena of inanimate bodies might be deduced. The *Principia* was the portion of this immense programme that Newton himself completed. In the Preface to it Newton wrote of his wider hopes when, after referring to his derivation of the celestial motions from the laws of gravitation, he continued :

> I wish we could derive the rest of the phenomena of Nature by the same kind of reasoning from mechanical principles, for I am induced by many reasons to suspect that they may all depend upon certain forces by which the particles of bodies, by some causes hitherto unknown, are either mutually impelled towards one another, and cohere in regular figures, or are repelled and

recede from one another. These forces being unknown, philosophers have hitherto attempted the search of Nature in vain ; but I hope the principles here laid down will afford some light either to this or some truer method of philosophy.

This passage alone renders it certain that the mechanical philosophy was no less the core of scientific understanding for Newton than it was for Boyle. For neither of them was it an artificial model or an empty hypothesis. It represented their assurance of the true structure of nature, which seemed to be vindicated by independent evidence at every turn. That philosophy alone set everything, from the walking of flies on water to the constancy of the stars in their courses, in the same coherent pattern of physical law.

Yet as both Boyle and Newton knew, experiments could only illustrate the truth of mechanism. They could not prove the existence of particles and intraparticulate forces. A view of nature as universal and deep-running as this could not in any significant sense be argued from " Experiments and Observations by Induction ", as Newton said a general scientific conclusion should be. On the other hand, it is difficult to imagine any practicable demonstration or experiment by which the tenets of the mechanical philosophy could be confuted in the seventeenth century. Was it, then, a theory that could be neither proved nor disproved ? Considered as a body of explanations of physical phenomena it was, just as the Copernican system of the universe was ; and a positivist philosopher might accordingly deny the importance of either. But it is a mistake to suppose that ideas that can neither be proved nor disproved are merely empty, as the history of Copernicanism shows. And so with the mechanical philosophy. Whereas the qualitative physics of Aristotle had led to sterility, the reduction to mechanical principles of those properties of bodies that were not mechanical in themselves gave promise of yielding quantitative and mathematical explanations. The strength of this philosophy was the fact that its ultimate realities, matter and motion, were ones that could be handled precisely by the existing science of mechanics. Whatever the weakness of existing proofs, there was nothing in principle intangible, inaccessible, incalculable or indeed beyond experiment in the final analysis that the mechanical philosophy hoped to attain. It was a philosophy of the concrete, the numerical and the measurable.

There has never been a more constructive change in scientific thought than that which brought it into favour. The views of Boyle and Newton were susceptible, without essential modification, of infinite elaboration and variation in their detailed applications. They asserted that the structure of the world was basically uniform and simple—and therefore understandable—but that from the permutation of this simplicity endless manifestations of different effects could appear. Science, therefore, was a unity in the form of its ultimate explanations of things, manifold in the paths it could trace back from things to their ultimate causation. To suppose this was, of course, to take a metaphysical position. Newton was not brought by the mechanical philosophy to declare, like so many others, that nature is ever simple and conformable to herself; rather it was his belief that it was so that made him a mechanical philosopher.

Among all the natural philosophers from Galileo to Newton, and after, the mechanistic outlook originated in an attitude of mind. Its first assumption—not entirely new in the seventeenth century but never before in the ascendant—was that a natural event was not to be accounted for by its relation to the divine power or some other mysterious universal principle; the correct procedure was to trace its antecedents and through these refer it to simpler circumstances whose mode of action was better understood. Such a procedure must end in the irreducible attributes of matter. Hence the second assumption was that everything happening in the universe—other than miracles—had a material cause; there were no immaterial agencies responsible for the course of events. Moreover, all matter was taken to be essentially alike—there was no matter privileged to be more active than the rest. The third assumption was that the simplest attributes of matter were its quantity (Newton's mass) and its motion. (Newton was compelled to complicate the metaphysic by postulating forces as causes of motion, but others argued that the starting and stopping of motion was a mere illusion, and hence that forces were needless.) The fourth assumption was that the uniformity of nature required that the quantities of matter and motion in the universe be unvarying.[2] The rest followed inevitably, for if the ultimate realities of matter and motion were conserved always, all change was a consequence of redistribution. And this in turn required parts that could move more or less, aggregate more or less, exert more or less force. The final, and for the direction of this attitude of mind specifically towards scientific problems, the major assumption was that the characteristics of these parts, or particles, could be inferred from those of gross bodies. As Newton put it :

We no other way know the extension of bodies than by our senses, nor do these reach it in all bodies ; but because we perceive extension in all that are sensible, therefore we ascribe it universally to all others also. That abundance of bodies are hard, we learn by experience ; and because the hardness of the whole arises from the hardness of the parts, we therefore justly infer the hardness of the undivided particles not only of the bodies we feel but of all others . . . The extension, hardness, impenetrability, mobility, and inertia of the whole result from the extension, hardness, impenetrability, mobility, and inertia of the parts ; and hence we conclude the least particles of all bodies to be also extended, and hard and impenetrable, and movable, and endowed with their proper inertia. And this is the foundation of all philosophy.[3]

That is to say, provided that the natural philosopher shares the attitude of mind that enables him to adopt these five assumptions (and to Newton and his contemporaries no others appeared possible), material particles and motion are given ; it remains only to discover how phenomena result from them. The assumptions people a stage with actors ; it is for science to give them speeches and gestures.

Immediately before the passage last quoted Newton wrote :
. . . since the qualities of bodies are known to us only by
experiments, we are to hold for universal all such as universally
agree with experiments ; and such as are not liable to diminution can never be quite taken away. We are certainly not to
relinquish the evidence of experiments for the sake of dreams
and fictions of our own devising ; nor are we to recede from
the analogy of Nature, which is wont to be simple, and
always consonant to itself.

For him, therefore, as for Boyle and perhaps Galileo, the metaphysics
that conduced to the mechanical philosophy was indissolubly one
with experimental science. Newton saw the metaphysical inference
that particles exist as indeed empty and useless by itself, until it
was given meaning by physical inference, from experiments, which
determined what the particles were and what they did. This is
the frontier that separates his epistemology from that of Descartes.
To put it more self-consciously than Newton would have done :
we do not gain understanding of the universe from metaphysics,
but we learn from them what form our understanding can take.
Understanding itself, however, comes from experimental and
mathematical science.

This was a way of looking at experimental science, raising it
above naive empiricism, that enhanced the importance of genuine
and exact experimentation. It regarded experiment as the arbiter
of what really is, for which only those experiments are valid that
have actually and carefully been made. It excluded appeal to
experiment as illustrative device, as Galileo had so often used it.
Boyle once commented adversely on some experiments in Pascal's
hydrostatics on the grounds that they could never have been
performed, as they involved men sitting at the bottom of deep
tanks and other improbabilities. No doubt Pascal's conclusions
from these thought-experiments were correct, but they were not
demonstrated to be so by such experiments. More than one
scientist objected, as Newton did, that mere thought-experiments

should not be brought up against ideas solidly founded on real ones ; or as Auzout complained, what right had Hooke to be so sure that his machine would grind improved lenses before he made it ? The true experimental scientist would be cautious of declaring with Galileo that the teachings of reason were so clear that experimentation was unnecessary.

Until somewhat after mid-century the main centre of experimental activity in physics was in pneumatics. Gilbert—following Norman and others—had temporarily exhausted the possibilities of experiment on magnetic and electrical bodies. Galileo and Stevin had done the same for hydrostatics. Much practical work was done on lenses and telescopes, and on various forms of thermometer as well (an intriguing device was Drebbel's thermostatically controlled furnace, one of the first feedback mechanisms), but all this was rather development of new instruments than experimental science. Mersenne in the 1630s, and the Accademia del Cimento later, made experimental tests of Galileo's laws of motion. Mersenne was also active with experiments on acoustics, again continuing Galileo's investigations. In the *Discourses* of 1638 Galileo had remarked upon the sympathetic vibration of strings, and with typical acuity observed that when a sharp steel chisel was pushed across a brass plate the distance between the lines scored diminished with the speed of the stroke and the pitch of the sound produced. Such a numerical coincidence was not to be ignored ; pursuing it, Galileo found that he could match his tuneful tool to two strings separated by a fifth. " Upon measuring the distance between the marks produced by the two scrapings, it was found that the space which contained 45 of one contained 30 of the other, which is precisely the ratio assigned to the fifth." * This observation led Galileo to affirm that the pitch of a sound is dependent upon its frequency, " the number of pulses of airwaves which strike the ear-drum, causing it also to vibrate with the same frequency ". Another experiment showed that

* The fifth is the harmony produced by striking two strings, one two-thirds the length of the other, that is as 30 to 45.

the number of standing waves on water in a "singing" wine-glass doubled if the note was made to jump an octave.[4] It seems also that Galileo determined or inferred the relative frequencies of fundamental, fourth, fifth and higher octave as 6, 8, 9 and 12.[5] And finally, he explained that notes of different pitch were harmonious to the ear when their successive "pulses" fell upon it at regular intervals, so periodically reinforcing each other.

Mersenne carried out experiments from which he discovered how pitch varied with the length, specific gravity, diameter and tension of the string. He also drew attention to the production of harmonics above the fundamental when an open string is struck, later accounted for by the independent vibration of the parts of the string between nodes. Mersenne's measure of the speed of sound, as about 1400 feet per second, from the interval between the flash and report of firearms, was reduced to about 1100 by subsequent experimenters.

Pneumatics, a far more controversial science, effectively began with Torricelli's barometric experiment (1643). Its origin, once more, lies with Galileo who had tried to account for the impossibility of raising water more than about thirty feet with a suction-pump. He thought that Nature's resistance to a vacuum (the so-called *fuga vacui*, usually invoked to explain phenomena of suction) was not unlimited, but fixed ; so that when the column of water was more than thirty feet long its weight disrupted the cohesion between the particles of water and the column broke like an overstrained wire.[6] To test the assertion, Gaspero Berti in Rome constructed a water-barometer with a tube over thirty feet long ; when unstopped at the base the water fell to the height stated by Galileo, leaving an "empty" space at the top. There was dispute among Berti's friends on whether or not the empty space was a vacuum, and on the cause of the water's suspension. Emmanuel Maignan and some others thought that the weight of the water was counterbalanced by the pressure of the atmosphere on the bottom of the tube ; the rest could not understand how the air could be said to exert a pressure on bodies immersed in it.* A few years later Torricelli and Viviani repeated the experiment using mercury instead of water, finding as they expected that the length supported was reduced in proportion to the greater density of mercury. Torricelli adopted Maignan's atmospheric interpretation of the effect. Indeed, one of his principal reasons for performing the experiment was to observe variations in the atmospheric pressure. He also attempted vainly to introduce insects into the empty space above the liquid. Torricelli's apparatus was seen by Mersenne on a visit to Italy, and after his return the experiment was repeated in France (1646). It was also performed, quite independently, by a Capuchin friar in Warsaw, Valerio Magni. Mersenne, however, gave permanent currency to Torricelli's name as that of the first inventor of the barometric experiment.

The related ideas that there is a vacuum above the fluid in a barometer, and that the fluid is supported by atmospheric pressure (or the weight of the air), were not rapidly accepted. Descartes, rejecting them altogether, declared that the Torricellian experiment would not succeed in an airtight room! Others held, as Galileo had done, that the vacuum exercised a positive attraction. Others still, Roberval among them, explained the effect by the conservative force of nature. At this time there was still, despite Stevin's earlier work, an almost universal confusion of ideas concerning the transmission of pressure in fluids, and some who attempted to calculate the height of the atmosphere from the supposed relative densities of air and mercury, in conjunction with Torricelli's experiment, agreed that the figure they obtained was much too small.

One who took up the experimental study of these questions in France was Blaise Pascal (1623–62). In a pamphlet of 1647 he described variants of the Torricellian experiment, and others made

* Before these events atmospheric pressure had been adduced by Isaac Beeckman and Giambattista Baliani to explain the ascent of water in suction-pumps, but this idea was by no means widely entertained.

with bellows, siphons and so on, combining to convince him of the reality of the vacuum :

I believe [he said] that the experiments recorded there suffice to show indubitably that nature can, and does, tolerate any amount of space empty of any of the substances that we are acquainted with, and that are perceptible to our senses.[7]

In the most conclusive experiment a barometer three feet high was enclosed in another six feet high. When the mercury was allowed to fall down in the longer tube, the shorter one hung in the empty space at the top—and the mercury in this fell down completely into its reservoir. When the seal at the top of the six-foot barometer was broken all its mercury ran out, but that in the small one rose again to the normal height. As yet, however, Pascal was not quite convinced that the *fuga vacui* was a chimera, though he suspected that the mercury was really supported by the " weight and pressure of the air because I consider them only as particular cases of a universal principle concerning the equilibrium of fluids ". He proposed to his brother-in-law Périer an experiment perhaps first suggested by Mersenne or Descartes : the ascent of a mountain with a barometer to see whether the mercury level fell as the weight of air above diminished. Périer lived at Clermont, in Auvergne, conveniently at the foot of the Puy-de-Dôme. Nearly a year later (19 September 1648) the portentous ascent was made ; the glass tube remained unbroken and when filled with mercury at the summit the level stood at 23·16 inches—3·12 inches under the level at the base 3000 feet below. Pascal's anticipation was triumphantly confirmed.

Thus reassured, Pascal renounced *fuga vacui* altogether. The report of the experiment won Pascal great fame, yet the experiment itself—which was not highly original or recondite—was really less impressive than the theoretical study that Pascal perfected in the next five years. Its significance is that it treats air as a fluid, like water ; Pascal made the principles of pneumatics identical with those of hydraulics, apart of course from the compressibility of air.

Admittedly every point made by Pascal in hydraulics had been established previously by Stevin, but Stevin's writings had had no wide audience. Pascal re-stated the uniformity of pressure in all directions in a liquid, and applied this principle to the atmosphere. He remarked that a mercury-siphon would work as well in water as in air ; moreover, if the siphon were deep under water, a tube could be led out from the top into the air without impairing its operation, thus neatly excluding the *fuga vacui* explanation.* It was undoubtedly his interest in hydraulics that induced Pascal to explore and extend the Torricellian phenomena, and it was the idea of the atmosphere as a " sea of air " that suggested the Puy-de-Dôme experiment. There is perhaps a certain irony in the last words of Pascal's treatise :

> Let all the disciples of Aristotle collect the profoundest writings of their master and of his commentators in order to account for these things by abhorrence of a vacuum if they can. If they cannot, let them learn that experiment is the true master that one must follow in Physics ; that the experiment made on mountains has overthrown the universal belief in nature's abhorrence of a vacuum, and given the world the knowledge, never more to be lost, that nature has no abhorrence of a vacuum, nor does anything to avoid it; and that the weight of the mass of the air is the true cause of all the effects hitherto ascribed to that imaginary cause.[8]

For the logic of Pascal's pneumatics was theoretical rather than empirical, and few of the experiments he alleged can ever, in fact, have been carried out.

A more practical treatment of air as a fluid was that of Otto von Guericke (1602–86), who discovered that it was possible to pump air as though it were water. Until he realised that the expedient was needless he had attempted to exhaust vessels filled with water in order to produce a vacuum mechanically. His famous demonstration of the force of atmospheric pressure—von Guericke rejected *fuga*

* However, Pascal could certainly never have made this difficult experiment.

vacui though his experiments were independent of the barometric ones—in the "Magdeburg Experiment" of 1654, when teams of horses failed to separate a pair of exhausted brass hemispheres about three feet in diameter, was the culmination of years of effort. The first account of it by Caspar Schott in *Mechanica Hydraulica Pneumatica* (1657) was read by Robert Boyle, who soon devoted his own experimental genius to exploiting the possibilities of the air-pump constructed by his able assistant, Robert Hooke.

The air-pump was the unfailing *pièce de résistance* of the incipient scientific laboratory. Its wonders were inevitably displayed whenever a grandee graced a scientific assembly with his presence. After the chemist's furnace and distillation apparatus it was the first large and expensive piece of equipment to be used in experimental science. The early form, in which a glass receiver was cemented to a brass plate and exhaustion effected by means of a rack-and-pinion pump with manually operated valves, remained essentially unchanged for a hundred and fifty years. Boyle used a barometer as gauge of the reduced pressure inside the receiver, and either he or Hooke contrived a removable lid for it as well as various devices working through an air-tight joint to permit manipulation inside. Thus a wide range of experiments became practicable. Boyle published three series of them, in 1660, 1669 and 1682 ; the first created his reputation as a scientist and was imitated everywhere.[9] This dealt chiefly with the physical effects of the vacuum, the expansion of a limp bladder, the rupture of thin glass, the descent of mercury in a barometer,* the extinction of a flame, the death of a bird and a mouse, the loss of transmission of sound, and so on. Boyle distinguished between the weight of the atmosphere and its elastic force, which was alone active in a sealed receiver and could expand its volume at least 200 times.

This notion [he wrote] may perhaps be somewhat explained by conceiving the air near the Earth to be such a heap of little

---

* With his first pump Boyle could not reduce the mercury below one inch ; with his second he claimed to remove the residual air pressure completely.

bodies, lying one upon another, as may be resembled to a fleece of wool. For this consists of many flexible and slender hairs, each of which will, like a spring, be still endeavouring to stretch itself again. For though both these hairs, and the aerial corpuscles to which we liken them, do easily yield to external pressures ; yet each of them (by virtue of its structure) is endowed with a power or principle of self-dilatation. . . .[10]

For his mechanical interpretations—not limited to this rather crude model—Boyle was criticised by Thomas Hobbes and Francis Linus. It was in the course of his reply, accompanying the second edition of *New Experiments Physico-Mechanical* (1662), that Boyle announced the relation, "that pressures and expansions be in reciprocal proportions" $(pv = k)$.[11] It is one of the oldest examples of a significant physical relationship's being gathered from tabulated experimental data, and for that alone was a triumph for the empirical method Boyle loved.

Despite improved technique, the *First Continuation* (1669) added few new experiments. The most interesting are those on capillary attraction, and those in which Boyle tried to discover whether or not there is resistance to motion in an exhausted vessel. Dropping a square of feathers through two feet of vacuum was an insufficient test. Boyle related how

being accidentally visited by that sagacious mathematician Dr [Christopher] Wren, and speaking to him of this matter, he was pleased with great dexterity as well as readiness to make me a little instrument of paper on which, when it was let fall, the resistance of the air [was very conspicuous].

Wren's handiwork was unfortunately lost at the time when the experiments were made. In the *Second Continuation* (1682) many of the trials were made by Denis Papin (1647–1712 ?), who had formerly assisted Huygens and has as good a claim as anyone to be considered the inventor of the internal combustion engine ; as they were largely devoted to exploring the " airs " yielded by fermenting organic matter they belonged rather to chemistry than to physics.

Twice, in introducing the successive *Continuations*, Boyle lamented the failure of others to follow his example in making original pneumatic experiments. The reproach was perhaps a little unjust, for not only were few scientists as well able as Boyle to employ workmen to make pumps and assistants to do routine work, but there was keen interest elsewhere, notably in the Accademia del Cimento. Yet it is true that the salient observations were all Boyle's, that he contrived the best justification for the mechanical interpretation of pneumatic phenomena, and that he showed how the air-pump could be utilised for a wide range of experiments. Boyle had a touch of the dilettante, deserting the scene if he was cold or out of sorts, bored or summoned by a visitor (while constantly complaining that incessant social occasions interrupted his work), and he could not always be bothered to work up his results and ideas properly ; for all that, in all his multifarious experimental researches his was always the inquiring and fertile mind. It was Boyle who made pneumatics, in the spirit of Baconian empiricism, a truly experimental branch of science rather than a topic for speculation and *a priori* reasoning.

Most of Boyle's experiments were repeated by the Royal Society after he had presented to it his first air-pump. At about the same time, in 1662, he also recommended to the Society the man who had made it, Robert Hooke, as Curator of Experiments. Some, with slight justification, have attempted to make him the leading figure in Boyle's physical experiments, but it could more reasonably be maintained that it was from Boyle that Hooke acquired sound principles of experimental science. The faculty that had served Boyle well, Hooke's mechanical ingenuity, was also his weakness ; it led him into an unremitting hunt for gadgets, some good, some bad. Endowed with an inexhaustible imagination both for experiment and theorisation, he lacked Boyle's capacity to pursue an idea relentlessly through argument and experiment, just as he lacked Newton's depth of thought and mathematical power. Of the three great English physical scientists of the later seventeenth century Hooke was the lightweight, too apt to promise what he could not always perform, too ready to offer a theory pat for every occasion. If he impressed contemporaries the more (like young Samuel Pepys, who wrote that Hooke promised least and performed most of any in the then infant Royal Society—but how many scientists did Pepys know then ?), his work has impressed posterity the less. Yet Hooke's talents, though not of the kind that transform science, were almost perfectly adapted for the service he did to the Royal Society. That body, many of whose members were individually more cautious than Hooke, wanted its curator to challenge the world with new experiments and demonstrations, and this he did superbly.

From its foundation until the early eighteenth century the Royal Society was the chief European centre of experimental physics. For although by no means all the significant experiments of the time were demonstrated to the Society at Gresham College— not even all those of such Fellows as Boyle and Newton*—a vast number of experiments was made at or in connection with its meetings. The chief actor was Hooke, but there were many other prominent performers, among them Boyle, Wren, Huygens and other foreigners on their visits to London, Brouncker, and Halley— whose ingenuity was not far below Hooke's. What was done was varied and incoherent, its results often inconclusive. There were the inevitable mechanical experiments, many on clocks and watches, others on the expansive force of gunpowder and on ballistics and, to balance this apparently utilitarian emphasis, a great deal of time was spent in the early days on "weighing the air", to Charles II's amusement. Hooke spoke of many new instruments and devices, some of which were actually constructed : the wheel-barometer, thermometers calibrated on various plans, sounding machines, an air-pump receiver large enough to hold a man (this never worked),

---

* Strangely enough it seems that Newton's famous experiments on the prism were never repeated before the Society. Newton is not known to have demonstrated any of his original experiments in public.

a philosophical lamp with a constant-level reservoir, a new navigational instrument, and so on. Other topics were of wider theoretical interest : the suggestion that the pendulum beating seconds be used as a standard of length—defeated by the discovery that it varied with latitude; attempts to detect variations of gravity above and below the surface of the Earth ; comparison of the theoretical laws of impact with experiments on colliding bodies. All in all it was a lively if haphazard programme, in which the testing of hypotheses and of new measuring instruments, the proof of new views and the disproof of old ones, the determination of physical constants and of astronomical positions, were all inextricably confused. Inevitably the fruits of such randomly directed collective energy were small in proportion to the effort and to those gathered by concentrated individual research. Something was done to refine the equipment of the scientific laboratory. The notion, and to some extent perhaps also the method, of pursuing experimental physics were more widely diffused. And the Society was established as guardian of the public conscience of science. If apparatus failed to work on a Wednesday afternoon it had to be brought in for a second or third trial. When Newton's electrical experiments could not be duplicated in London the Society badgered him for further details and instructions until they did—though rather less dramatically than Newton's account had led them to believe. Nor was the Society always willing to take the investigator's word for the interpretation of an experiment, even when the facts were verified as correct.

The classic instance is the reception given to Newton's epoch-making letter on light and colours, read to the Society on 8 February 1672. It received favourably this first communication from a Fellow of one month's standing, committing it to Seth Ward, Bishop of Salisbury (a mathematician), Boyle and Hooke, " to peruse and consider, and bring in a report of it to the Society ". Apparently the Bishop and Boyle never did report formally, but Hooke was back in a week, embarrassing his colleagues with " so sudden a refutation of a discourse . . . which had met with so much applause at the Society but a few days before ". Seven years senior to Newton in age, ten years in the Fellowship, Hooke was the Society's authority on experimental physics and had put forward his own theory of light in *Micrographia* (1665). He was a strong partisan of the view that light is a periodic motion, and he had convinced himself that Newton was not. He set out to demolish Newton's " hypothesis " (p. 126).[12]

The problem tackled by Hooke and Newton—the appearance of colours when light is refracted—was of millennial antiquity. A good step forward was taken when in the Middle Ages the rainbow colours were definitely associated with this bending of light as it passes from one medium to another, by Alhazen in Islam and Theoderic of Freiburg in the West. Thenceforward, however, there was no experimental, theoretical or geometrical advance in the optics of refraction until towards the end of the sixteenth century, and indeed some of what had been accomplished was lost. The geometry of refraction was settled empirically by Willebrord Snel in 1621 and made generally known by Descartes (who may or may not have discovered Snel's Law independently) in 1637. This did not explain the formation of colours, a physical problem. Descartes went even further : repeating medieval experiments on the passage of rays of light through globes of water, with the aid of Snel's Law he explained the geometry of the rainbow arc (approximately) in *Les Météores*, but this still did not explain the colours of the bow. Nevertheless, Descartes was not without a theory ; the particles of the matter of light, he declared, pressed outwards from the centre of the vortex, or a flame, must spin as well as travel, a fast spin being perceived as red, a moderate spin as yellow, a slow spin as blue. Thus the formation of colour was a mechanical change in the matter of light, one which it acquired (for instance) by its oblique impact on the surface between two media causing the particles (like tennis-balls striking the ground) to begin to spin. Incidentally, Descartes' theory of light was quite inconsistent with itself, for at the cosmic level he made light an instantaneously transmitted pressure, whereas at the physical level he made the particles move progressively to provide an explanation of refraction.

In contrast to Descartes' tennis-ball picture Hooke's theory ignored the texture of the transparent medium through which light moves to concentrate on its motion, which he regarded as vibratory. Light, he thought, consisted of a succession of spherical pulses generated by the source, exactly like the pulses of sound, which in a narrow beam travelled at very minute intervals apart as planes perpendicular to the beam. Should such a beam fall upon water (out of air), which Hooke like Descartes supposed to transmit light more easily than air, each pulse in the train passing through the interface would be for an instant less resisted on one side than on the other. This would have the effect of making the pulse oblique to the beam, until it was again equally and oppositely refracted. It was this obliquity that affected the eye as colour. The ray A in the beam (figure 10) has its " leading " edge bruised or weakened by the unilluminated medium ; the ray B, however, has its " lagging " edge thus weakened. Hence,

> we may collect these short Definitions of Colours : That Blue is an impression on the Retina of an oblique and confus'd pulse of light, whose weakest part precedes, and whose strongest follows. And, that Red is an impression on the Retina of an oblique and confus'd pulse of light, whose strongest part precedes, and whose weakest follows.[13]

As for the more fortunate middle rays of the beam, they produce middling colours like yellow and green. Although Hooke remarked that a mixture of red and blue pigments gave purple, he supposed that mixture of the two ends of the spectrum gave green, for this reason the most grateful colour to the eye. Nor did he notice, in all his " hundreds " of experiments, that it is violet not deep blue that terminates the spectrum.

Hooke did not describe any experiments on the refraction of light by water in detail, but he did apply his ingenious theory to double

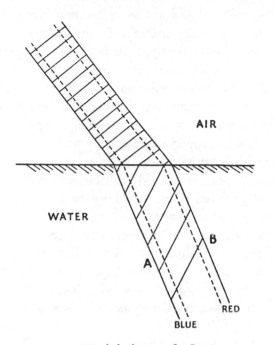

FIG. 10. Hooke's theory of refraction

refraction and a single reflection in a globe to show that the observed order of colours in the rainbow results. His observations on the colours of thin translucent plates, for which he sometimes used his microscope, were entirely novel. In thin sheets of mica the colours were arranged in rings about a central coloured spot, in the order of the secondary rainbow. Occasionally patches or streaks of colour were visible. The same colours appeared when two pieces of glass were pressed together, even in a very thin single piece, and in bubbles of all sorts. Hooke noted that a certain thinness was required, and that below this limit the colour varied with the thinness of the layer through which the light passed. He attributed a variety of natural coloration, like that of mother-of-pearl, to this effect. But he was wrong in supposing that the colours could be seen only by reflected and not by transmitted light, a notion that formed the basis of his explanation. For he reasoned that some light would be reflected from the upper surface of the thin plate, and some —weakened by two refractions—from the lower surface ; thus there would be two sets of pulses in the resultant reflected beam, the one stronger, the other weaker, the interval between them varying with the thickness of the plate. Considering any pair of these pulses, unless they should happen to be equally spaced in the train either the weaker will just precede the stronger, or just follow it. Thus a colour would appear, in accord with Hooke's earlier definition of colour ; and coloured rings would appear as the thickness of the doubly reflecting plate varied.

In Hooke's view, as in Descartes', therefore, coloured light was a modified form of white light. Red, white and blue were three distinct physical states. Consequently, by reversing the modification, any one of these could be turned into any other. Hooke was always convinced that his theory completely sufficed to explain all the phenomena of light and colours : it was eminently mechanical, it fitted the facts, therefore it must be true. Like Descartes, Hooke sometimes could not help believing that what is ingenious must really exist, and his theory of light was certainly that. In his near approach to the concept of interference it was more ingenious than Newton's.

Newton was probably engaged on his first experiments with the prism in 1664 before *Micrographia* appeared. At any rate when he read *Micrographia* he knew enough to express forcibly his dissent from Hooke's views. Probably the essential series of experiments— in which he seems to have rediscovered interference phenomena— was complete by the end of 1666. But six more years passed, and Newton had already delivered three brief courses of lectures on optics as Lucasian professor of mathematics at Cambridge, before his work was made known to the Royal Society, already the recipient of Newton's reflecting telescope. They involved, he wrote, a Philosophicall discovery wch induced mee to the making of the said Telescope, & wch I doubt not but will prove much more gratefull then the communication of that instrument, being in my Judgment the oddest if not the most considerable detection wch hath hitherto beene made in the operations of Nature.[14]

No slight claim !

Characteristically, Newton chose for his major investigation the simplest form of refraction, occurring when light shines through a triangular glass prism. With a good sense not always shown by earlier experimenters he made his observations in a completely darkened room, and with a narrow beam of light. He placed the screen on which the refracted light fell at a distance from the prism, so that the spectrum was large and full, not a white patch surrounded by coloured fringes. By training and temperament a mathematician whose thought tended towards quantitative expressions, he was prepared to see not merely an unexpected effect, but its significance. Newton expected to see a round image corresponding to the sun's disk. Instead he saw a roughly oblong spectrum with its axis at right angles to that of the prism. " Comparing the length of this coloured Spectrum with its breadth ", he wrote, " I found it about five times greater, a disproportion so extravagant, that it excited me to a more than ordinary curiosity of examining from whence it might proceed." It made no difference whether the beam penetrated the prism near its apex or its base, nor did twisting it to and fro slightly shift the spectrum. From Snel's Law Newton computed that the breadth of the spectrum was correct, but the law did not yield anything like its angular length of 2° 49′. And certainly the refracted beam was not curved, since the length of the spectrum was always in proportion to its distance from the prism.

It seemed, then, that the different coloured rays must diverge from their origin, and that one ray must be more markedly bent than another. To test this hypothesis Newton in his next trial (which, echoing Bacon, he called a crucial experiment) shone the white beam of sunlight through one prism, then through two holes in boards to make a precise, narrow beam, and finally through a second, fixed prism. By twisting the first prism slightly he could make the narrow beam, shone through the second prism, of any colour. Observing the positions of the images on a screen, he found that they varied with the colour of the light reaching the second prism : the blue rays were more sharply refracted than the red.

So apparently simple was Newton's discovery of the dispersion of light by refraction that it deceived many of his contemporaries. For Newton himself it was only the beginning of his reasoning on the nature of light, which he supported by still further experiments. First, he left off his attempts to improve lenses, his " Glassworks " :

> for I saw, that the perfection of Telescopes was hitherto limited, not so much for want of glasses truly figured according to the prescription of Optick Authors, (which all men have hitherto imagined), as because that Light it self is a Heterogeneous mixture of differently refrangible rays.

With this decision—a mistaken one—Newton set a course which was temporarily adverse to the development of optical instruments. Spherical aberration was not so trivial an imperfection as he supposed, nor was chromatic aberration so far beyond correction as he imagined. But the same decision initiated his fruitful trial of a reflecting telescope. In the second place, and much more important, he went on to prove that white light is a " confused aggregate of Rays indued with all sorts of Colors, as they are promiscuously darted from the various parts of luminous bodies ". Each ray had its characteristic colour and degree of refraction, neither being mutable by any means. Thus a primary green from the prism could not be altered, although a composite green—made by mixing primary yellow and blue light—could be again divided into its elements. The prism was a filter (one might say) that divided white light into its parts by refraction ; and similarly in reflection the surfaces of bodies split white light, appearing coloured because " they are variously qualified to reflect one sort of light in greater plenty than another ". Hence, said Newton supporting Boyle, colours could not be considered qualities of bodies.

In his letter to the Royal Society Newton only explained briefly what his further experiments were, though he had described them in his optical lectures. In one, light from a prism, dispersed into coloured rays, was collected by a large convex lens and brought to a focus, where it was again white. A little short of the focus the spectral colours appeared on the screen ; a little beyond it they reappeared in reverse order. Here Newton clearly felt that he could virtually *see* the coloured rays analysed out of, and recombined into, white light. Furthermore, when any coloured ray was prevented from reaching the lens, the focal image was coloured by the mixture of the rest. If the interruption was rapid, persistence of vision—a physiological phenomenon Newton studied with some care—caused the image to appear always white. All the experiments indicated that normal white light was invariably composed of the full complement of colours, and that no pure ray from the spectrum could ever be modified into anything else. Moreover, each ray was mathematically labelled by its unique angle of refraction, which sufficed to explain many effects, notably the constant width of the rainbow.

At first Newton answered his critics with patience, especially in so far as their objections arose from misunderstanding of his experimental procedures. Later he felt he had become a slave to philosophy, bound to the treadmill of grinding out answers to all and sundry. His policy of interpreting experimental results not merely plausibly (hypothetically, as he would have said) but rigorously was hard to comprehend. In a passage suppressed in print Newton had written :

A naturalist would scarce expect to see ye science of [colours] become mathematicall, & yet I dare affirm that there is as much certainty in it as in any other part of Opticks [that is, geometrical optics]. For what I shall tell concerning them is not an Hypothesis, but most rigid consequence, not conjectured by barely inferring 'tis thus because not otherwise or because it satisfies all phaenomena (the Philosophers universall Topick) but evinced by ye mediation of experiments concluding directly & without any suspicion of doubt.[15]

None of the critics to whom Newton first responded touched the mathematical structure of his doctrine of light, on which he could confidently rely. To the extent that they attacked the method of his reasoning it was enough for him to announce his position for the first time :

If everyone is allowed to guess at the truth of things from the mere possibility of hypotheses, I do not see how anything can be established with any certainty in any science, since one can always think up more and more hypotheses and furnish fresh difficulties.[16]

And Huygens' two-colour theory of light could be exploded by pointing out that white light made from a mixture of yellow and blue would be analysed into yellow and blue by a prism, not into the full spectrum *de facto* derived from natural white light. (Here Newton was on the verge of distinguishing between the physics of colour and the physiology of colour perception, thoroughly confused by his contemporaries.)[17]

The reply to Hooke was a different matter. For Newton the coloured rays in white light were like straight fibres in a strand of silk, each a geometric and physical entity existing in the white beam, not created by refraction, nor fragments of a formerly homogeneous substance or motion—for he did not say how light was to be understood mechanically. That separation, not transformation, occurred in refraction was the hardest point of all in Newton's doctrine to prove ; Hooke challenged it :

Why there is a necessity, that all those motions, or whatever else it be that makes colours, should be originally in the simple rays of light, I do not yet understand the necessity, no more than that all those sounds must be in the air of the bellows, which are afterwards heard to issue from the organ-pipes. . . .[18]

One can see Hooke's difficulty. If each colour of the spectrum were physically distinct from the rest (as Newton supposed), how could the physical identity of each be maintained when all were mingled in white light ? It was simpler to suppose that the colours were not merely revealed, but created by refraction. Hooke never realised the importance of the point that each light-ray has besides a physical property (colour) a *mathematical* property (the specific degree of refraction associated with each colour). Therefore, in the first place, each colour was not just an "impression on the retina" of the eye ; it was something mathematically definite that even a colour-blind man could recognise. And secondly, this mathematical individuality of each colour could not be created by a qualitative transformation of white light by refraction, as Hooke supposed. It must exist in the white light before refraction. Consequently, Newton argued, white light must be composed of heterogeneous coloured rays. He could state this because he had, as it were, labelled each ray by its characteristic refraction.

In the last analysis Hooke's theory failed because it was geometrically amorphous. The modification of white light that it postulated was qualitative, not one linked to the geometry of the rays. It did not account for the great extension of the length of the spectrum. It could not explain the fact that when light was shone first through one prism, then through a second at right angles to the first (figure 11), the image formed was neither round and white, nor square and coloured, but a spectrum as before twisted through 45 degrees. (Newton also devised other experiments in which an apparently white beam, without any further refraction, was made to yield coloured images ; here was the strongest proof that the colours were really latent in the white.) Newton's theory explained all such experiments because in it everything followed from the geometry, each ray (with its own geometric individuality settled by its own constant of refrangibility) obeying Snel's Law.

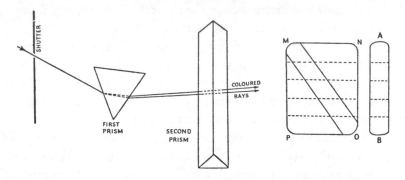

FIG. II. Refraction through crossed prisms. AB, image formed by first prism alone. MO, oblique image formed by crossed prisms. MNOP, image that would be formed if each ray were scattered and spread in two directions by the two prisms

Thus geometry dictated that white light be heterogeneous, not homogeneous.

Just as happened in the theory of gravitation, Newton's geometrisation of physics empowered him to transcend Hooke. But the very argument from the mathematical uniqueness of each coloured ray that refuted Hooke's scepticism was also responsible for Newton's mistaken view that chromatic aberration was irremediable, in much the same way that Galileo's geometrisation of physics led him into the error of circular inertia. For Newton always conceived of the different refrangibilities of the coloured rays as unchanging properties of those rays ; that is to say, dispersion was for him a property of light, not of the glass or other refracting medium, because dispersion was caused by the unequal refrangibility of the rays. Since this did not alter with one kind of glass or another, he was sure that whatever the material of the lens, the dispersion would be in a constant ratio to the refraction, and checked his confidence by some rather casual experiments. In this instance Newton was wrong, because dispersion and refraction are distinct properties, and the relation between them does vary with the type

268

of glass; as Chester Moor Hall showed empirically in 1733, achromatic lenses can be made.

Newton's letter of 1672 contained a promise to account for the colours of thin plates which he redeemed three years later. The real strength of Hooke's optical theory in *Micrographia* had been that it related the colour to the distance between the two surfaces of the plate by considering the periodicity of the light-pulses. Now Newton was obliged to reintroduce a similar notion, rather disingenuously remarking that "this, I suppose, they will think an allowable supposition, who have been inclined to suspect, that these vibrations might themselves be light". He prefaced his paper with a clear indication that this time he dealt with a hypothesis "because I have observed the heads of some great virtuosos to run much upon hypotheses", and that these were not to be confused with the demonstrable truths of the earlier letter. The foundation of Newton's speculation was an aether, in which the impact of light on bodies caused vibrations or waves to form. These spread faster than light itself. If the ray of light fell on a transparent body at the same moment as the crest of an aether-wave, it would be reflected; if at the same moment as a trough between two aether-waves it would pass into it. Thus, using homogeneous coloured light, when a convex glass lens was pressed hard on a flat glass plate rings of coloured light could be seen by transmission on one side, and alternate rings by reflection on the other, the intervals between the rings depending on the curvature of the glass and the air-space between (figure 12).

With white light the rings became tinged with the spectral colours, in the order violet (towards the centre), blue, green, yellow, red (outermost), repeated again through the next series of rings. Newton speculated that because the red-making gap of air was larger than the violet-making gap, the aether-waves accompanying red were longer than those accompanying blue, with the other colours in between. By the wavelength of the accompanying vibrations, white light was divided into colours through its

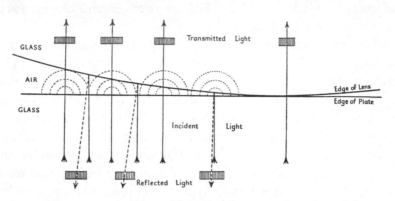

FIG. 12. Newton's theory of " fits ". The semicircles indicate " fits of transmission ". When the incident light-ray falling on the lens coincides with one of these it is transmitted into the glass. Half-way between these " fits " are the " fits of reflection " (not drawn) ; when the incident ray coincides with one of these it is reflected back from the surface of the lens. The " fits " shown belong to one colour only. The other colours would fall between, according to the length of their " fits "

propensity to be reflected or transmitted, just as it was divided in refraction by its varying refrangibility. And again as with refraction, the analysis of white light into its coloured constituents could be represented geometrically, though this time the geometry was more complex. For here the appearance of the coloured rings depended both on the shape of the thin plate and on the wavelengths associated with each colour. Newton made many experiments (described in his letter of 1675, and repeated in *Opticks* in 1704) to elucidate the phenomena of these rings of light, and found that his geometrical hypothesis could account for all of them :

> as all those things follow from the properties of light by a mathematical way of reasoning, so the truth of them may be manifested by Experiments.

For now Newton could abandon his aetherial hypothesis ; it was necessary only for him to postulate, as he did in *Opticks*, that the rays were put into alternate regular "fits" of easy transmission and easy reflection, without seeking a physical explanation. From this postulate onwards the rest of the theory could be mathematical. Newton's most remarkable achievement in it was to calculate from the dimensions of his apparatus the sizes of the air-gaps at which the various reflected rings were found ; they were successively as $5\frac{1}{2}$ millionths of an inch multiplied by the series of odd numbers ($5\frac{1}{2}$ millionths of an inch is actually about one-quarter of the wavelength of yellow light). Now since Newton's theory required that the rays be reflected when the width of the thin plate (or air-gap) was an odd multiple of half a wavelength of the "fits", it followed that this length was about eleven millionths of an inch for yellow light—approximately half the modern value.[19]

Evidently Newton's theory of light was very far from being a simple corpuscular or emission theory. The wave-concept was always essential to it—not as a hypothesis, but as a feature of a mathematical theory from which verifiable predictions could be drawn. Yet Newton always denied that this wave- or pulse-motion constituted the very transmission of light itself. From the beginning he believed that light must be in some sense material, though careful to insist that as far as his own discoveries were concerned this idea was no part of the mathematical laws which were in themselves quite adequate to account for the phenomena of optics. The fact that light travels in straight lines was for him decisive of this question : it suggested that light was emitted, and proved that light could not consist of waves since waves always bend round obstacles, as sound-waves and water-waves do. Yet curiously enough experiments showing that light does bend into the shadow were perfectly familiar to him. They were first described by Francesco Grimaldi (1613–63) in a posthumous book of 1665, so crucial a year in the history of optics. Grimaldi, professor of mathematics at Bologna, was one of the most active experimenters on light of the immediately pre-Newtonian age ; he observed that when a very narrow beam shone on a thin opaque object the shadow was wider than geometry predicted, and that when shone through little holes the spot was also bigger. In both cases shadow or spot were surrounded by coloured fringes. Such experiments were repeated by Newton, who concluded that light was inflected by bodies through which it did not pass, violet light (most refrangible) being least inflected and red (least refrangible) most inflected. Newton imagined that inflection, refraction and reflection were all, fundamentally, manifestations of the same power of bodies to act upon light.

Grimaldi, like Hooke, had believed that light was a vibratory motion but it was not until 1690, in Christiaan Huygens' *Traité de la Lumière*, that the undulatory hypothesis was given a precise geometrical expression. He supposed that light travels through a particulate aether, the vibrations of each aether-particle spreading a wave disturbance through its neighbours. When a large number of such wavelets reinforce each other a wavefront is created, the movement of which constitutes the radiation of light. Huygens was able to show geometrically that a wavefront so created would be refracted and reflected as a beam of light is (figure 13, a) ; however, he did not attribute any precise periodicity to his waves nor did he attempt to account for colour. Again, geometry proved that if (as Huygens supposed) each point of the wavefront acts as a centre for wavelets from which a new wavefront arises, light will only pass through an aperture in a beam without spreading markedly into the shadow (figure 13, b). He tackled also the difficult problem of double refraction in Icelandic spar (calcite), discovered by Erasmus Bartholinus (1670). Calcite formed double images, one of which was a product of rays of light recalcitrant to Snel's Law. Huygens dealt with these "extraordinary" rays by assigning to them a wavefront of a spherical curvature. But he could not explain the further effect (of polarisation) that he discovered when one crystal of calcite was placed upon another, for now the two types of ray seemed to behave asymmetrically. He abandoned this problem as insoluble.

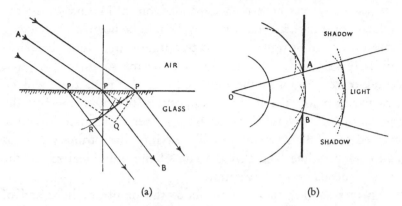

(a) (b)

FIG. 13. (a) Refraction of wave-motion. Huygens supposed that
light travels less quickly in glass than in air ; in a given time the
wavelets from every point P travel a shorter distance into the glass
than they would in air. Hence, instead of forming the wavefront
PQ they form a new one, PR. Geometry gives the displacement of the
refracted ray in accord with Snel's Law. (b) As the wavefront from
the light source o travels through the aperture AB only a small
number of wavelets cause an imperceptible light in the shadow on
either side of the beam

Newton returned to it in the *Quaeries* he added to the first Latin
edition of *Opticks* in 1706. He was unable to explain double
refraction completely by geometry, though his suggestion for an
explanation is of a geometrical kind. He judged from experiments
that light was not altered by calcite, any more than it was by
refraction or inflection. For the ordinary ray could be extraordinarily refracted by a second crystal, and vice versa. Hence
double refraction revealed a hitherto unsuspected property of light,
always present in its rays. And here Newton made a guess of
extreme insight, based on his sense of the fundamental geometry
of things. Every ray of light, he said,

may be considered as having four Sides or Quarters, two of
which opposite to one another incline the Ray to be refracted
after the unusual manner . . . and the other two . . . do not
incline it to be otherwise refracted than after the usual manner.[20]

How each ray was bent then depended on its orientation to the crystal. Some rays, inclined one way, would be normally refracted; the rest, inclined at right angles to the first group, would be extraordinarily refracted. If now the two beams so formed enter a second crystal arranged the same way as the first, they will be refracted as before ; but if it is oppositely placed to the first the orientations are reversed, and so the manner of refraction will be reversed too. This was almost like saying that ordinary light is randomly polarised, and that the fate of the individual ray in the crystal depends on its polarisation.

There are other speculations about the nature of light and of its interaction with material bodies in the *Quaeries*, but these do not detract from Newton's success in adhering to his design, which was as he expressly stated in the first sentence of *Opticks* " not to explain the Properties of Light by Hypotheses, but to propose and prove them by Reason and Experiments ". Reason meant, naturally, above all mathematical reasoning. Phenomena were to Newton adequately if not completely accounted for when they were traced back to universal mathematical laws, like that he postulated of fits of easy transmission and reflection. All the past problems concerning colours had been resolved by his apparently simple discovery that colour is a mathematically definite property, defined by the refrangibility of the coloured ray. All the discussion of colour as a form, or as a mechanical modification of uniform white light, or as a mixture of whiteness and darkness (as even Boyle and Barrow had supposed) had been mistaken and beside the point, since the need for it was utterly removed by the enunciation of that mathematical relationship :

> all the Productions and Appearances of Colours in the World are derived . . . only from the various mixtures or separations of Rays by virtue of their different Refrangibility or Reflexibility. And in this respect the Science of Colours becomes a Speculation as truly Mathematical as any other part of Opticks.[21]

For conventional optics had always been a geometrical science. The object of experiment was discovery of the mathematical relationships latent in nature, and confirmation of them by matching events with predictions, because the architecture of nature—and hence of science—is geometrical. *Opticks* could have been called after Newton's earlier and greater book, *The Mathematical Principles of Light and Colour*, since that is its subject. Newton in this lesser scope, the one he most fully explored in his own experimental researches, spoke in the same mathematical language of nature that he had employed in the *Principia*.

CHAPTER X

NEWTON AND THE WORLD
OF LAW

It is indeed a matter of great difficulty to discover, and effectually to distinguish, the true motions of particular bodies from the apparent ; because the parts of that immovable space, in which those motions are performed, do by no means come under the observation of our senses. Yet the thing is not altogether desperate ; for we have some arguments to guide us, partly from the apparent motions, which are the differences of the true motions ; partly from the forces, which are the causes and effects of the true motions. . . But how we are to obtain the true motions from their causes, effects, and apparent differences, and the converse, shall be explained more at large in the following treatise. For to this end it was that I composed it.[1]

By distinguishing the real from the apparent motion of falling bodies Galileo removed a principal objection against the Copernican system ; by discerning the beautiful simplicity of elliptical orbits behind the apparent complexity of the planetary revolutions Kepler increased the likelihood of its truth. Newton's achievement was founded on theirs, for by perfecting Galilean mechanics he found the basis for a final distinction between reality and appearance that placed Kepler's astronomy beyond doubt. The two great strands of seventeenth-century physical science were united in Newton's *Principia* (1687) through the fertile and fitting combination of mathematics with the mechanical philosophy that yielded the laws of gravity, when Kepler's planetary laws were shown to be but special cases of the laws of motion foreshadowed by Galileo. Four modes of thought represented by Newton's great predecessors Galileo, Kepler, Descartes and Huygens now appeared not merely harmonious but inseparable, so that by Newton's triumph the whole seventeenth-century revolution in science was confirmed.

Yet when Isaac Newton was born in 1642, at the turning-point of that revolution, the work of Galileo was still to be accepted, the discoveries of Kepler were still ignored, and the impact of Descartes, Boyle and Huygens was still to be felt. During Newton's boyhood science moved decisively into a new age. New methods based on mathematics and experiment rose rapidly to ascendancy, while new theories buoyantly exemplified the mechanistic, corpuscularian metaphysic. In England the Royal Society, a galaxy of talent, offered an appropriate setting for Newton's transcendent genius ; his intellectual inheritance was equally rich in critical problems and the means for solving them.

An outworn pattern characterised the education he received at school and university, and Newton rising from the lesser gentry lacked the social advantages of such grandees of science as Huygens and Boyle. Yet when he went into residence at Trinity College, Cambridge, in 1661 Newton had already acquired a practical mastery of Latin—still the international language of science—and probably of elementary geometry. In his first years there he studied Greek, logic, ethics, rhetoric (which meant the art of literary composition) and more mathematics. Somehow he also acquired a stumbling knowledge of French, though he never set foot abroad. From 1663 his unique gifts began to be conspicuous ; a couple of years' work enabled the pupil of Isaac Barrow (a considerable mathematician who had returned recently to Cambridge as the first Lucasian professor) to outstrip his master. At the same time he was pursuing physical science in the writings of Galileo, Kepler, Descartes, Boyle and many others. Newton was a late developer ; entering the university at eighteen he was three or four years older than many undergraduates, but in his student days he completed a thorough training in science and mathematics largely additional to his formal studies. In the next two years—1665-6—all his future accomplishments took shape. For the most part he spent them at his home at Woolsthorpe in Lincolnshire, for the university was closed by the great plague ; Newton's creative life as a scientist began on the farm. It was to continue through normal academic grooves. In the spring of 1667 he returned to his College, to be elected into a Fellowship and reside there for nearly thirty years. His small patrimony relieved him of the fear of academic poverty. Most dons were driven from their High Tables by boredom and the desire for a wife ; Newton was neither bored nor married and by a special dispensation from the Crown, given five years after his succession to Barrow in the Lucasian chair (1669), he was allowed to retain his Fellowship without the usual necessity for taking Holy Orders. The way was clear for science.

Newton wrote of his active rustication many years later :

In the beginning of the year 1665 I found the Method of approximating series & the rule for reducing any dignity of any Binomial into such a series. The same year in May I found the method of Tangents of Gregory and Slusius, & in November had the direct method of fluxions, & the next year in January had the Theory of Colours & in May following I had entrance into ye inverse method of fluxions. And the same year I began to think of gravity extending to ye orb of the Moon, & (having found out how to estimate the force with which a globe revolving within a sphere presses the surface of the sphere), from Kepler's rule ... I deduced that the forces which keep the Planets in their Orbs must [be] reciprocally as the squares of their distances from the centers about which they revolve : and thereby compared the force requisite to keep the Moon in her Orb with the force of gravity at the surface of the earth & found them answer pretty nearly. All this was in the two plague years 1665 & 1666 for in those days I was in the prime of my age for invention and minded Mathematics and Philosophy more than at any time since.[2]

There are, moreover, among Newton's notebooks and papers many records of optical experiments and mathematical calculations that belong to this early period, when Newton had only just ceased to be a student. One of the early experiments is that of viewing through a prism a piece of paper brightly painted with red and blue (which Newton described in *Opticks* forty years later), when the edges of the two coloured sections seemed out of line : " soe y^t blew rays suffer a greater refraction y^n red ones ", as Newton commented. He also composed a fairly full account of his new method of handling algebraic equations, and made notes on mechanics which show him discovering the law of centrifugal force and applying it to work out the strength of the Earth's pull on the moon. On his return to the university Newton was not short of ideas, nor matters to study further.

During the next few years, in fact, he gave most of his time to optics and mathematics. Presumably he continued his experiments with the prism and the colours of thin plates and almost certainly he was already drafting accounts of his mathematical discoveries and of his new theory of colours. Newton was always busy with the pen, either taking notes from his reading (with which he covered thousands of sheets), sketching out his ideas or making calculations. From January 1670 he lectured for about twenty years—on optics, mathematics, and in the last series mechanics. The burden was light, as it consisted of giving only eight lectures in the year, but Newton's were almost wholly written from original material and, indeed, went into his books in one form or another. At about the time that he succeeded Barrow, Newton's correspondence began to be extensive ; through one friend in London, John Collins (1625–83), his mathematical ability became fairly widely known, just as his optics did after his Royal Society letters of 1672. Newton had also developed a keen interest in chemistry, and there were periods during all his residence in Cambridge after the first few years when he was in constant attendance upon the furnace he set up in the little garden below his rooms.

Yet it is odd that in the record of two-thirds of Newton's life in Cambridge there is so little trace of interest in the problem he was to make peculiarly his own—the problem of gravitation. None of his friends was aware that Newton had more than a passing interest in mechanics and astronomy, and practically no hint of his thoughts of 1666 escaped him in any letter.* For fifteen years Newton pushed to the back of his mind one of the greatest of all scientific ideas because (apparently) it struck him as no more than a hypothesis and he thought he had better things to do than try to make more of it. He felt that he could not demonstrate his ideas of 1666 to his own mathematical satisfaction, and yet when he returned to them again in 1679 and 1684–5 the solution of the mathematical difficulties only took him a few months.

The questions in mathematical astronomy that Newton had considered at Woolsthorpe in 1666 he inherited from Descartes and Galileo. Ever since Tycho had abolished the rigid celestial orbs ("absurd and monstrous" Kepler had called them) the system of sun and planets had lacked cohesion. Plainly the planets did not wander erratically, but there was nothing to keep them fixed in their courses. Descartes had first perceived the need for a celestial balance of force, a sort of cosmic string that would prevent the planets escaping out of the solar vortex under the force of its rotation, and had sought to make the aether provide it. But he made no calculations—a Babylonian stargazer might have protested against the freedom of his assumptions. In Descartes' time there was something more serious to take into account : Kepler's laws of planetary motion. If Descartes ever isolated Kepler's theorems from the otherwise (to him) repellent farrago of weird notions in Kepler's books he paid no attention to them. In which he was no more unjust than virtually all his contemporaries.

There was another problem still. When Copernicus moved the

---

* Newton referred obliquely to his early calculations on gravity in a letter to Huygens of 1673. For some unknown reason the passage was dropped from the copy of the letter received by Huygens ; did Newton reflect that it might provoke Huygens to think too successfully of the connection between dynamics and astronomy ?

Earth he knocked the bottom out of the old doctrine of lightness and heaviness ; if bodies continued to fall they did so without any rational reason for their behaviour. Copernicus had put forward the rather vague notion that pieces of Earth or Venus or Jupiter had a tendency to run together like drops of water. William Gilbert, being a student of magnetism, correlated this notion of heaviness with magnetic attraction, which of course he found in the Earth. The attraction of Earth for earth and so on was quite specific, however, since otherwise (as Aristotle had argued long before) all the matter in the universe, it seemed, would congregate in one vast unhappy heap. Galileo successfully cleared up the difficulty about lightness by denying that there was any such thing, all bodies being absolutely heavy and only relatively light ; but the actual cause or nature of this universal heaviness he left severely alone, apart from some remarks indicating that he found Gilbert's statements plausible.

Nor, while much concerned with the dynamical effects of heaviness, did Galileo contemplate at all the question of what it is that holds the solar system together. His concept of circular inertia relieved him from that necessity. It was Kepler who began to associate the problem of heaviness with that of the cohesion of the heavens, in his own highly eccentric fashion. He maintained that like matter would always be drawn to like, the distance each fragment moved being proportional to its " body " (mass). Thus, he wrote in 1609, the Earth would drag the moon down to itself, while rising a little way to meet it, if the two bodies were not restrained in their orbits by their own " animal or other equivalent forces ". He estimated quantitatively the tendency towards a motion that never actually happens and pointed to the moon's influence over the tides as evidence for the reality of mutual attraction ; like Descartes later, however, Kepler did not further explore his idea dynamically. He was no less blind to the astronomical portent of the science of motion than Galileo was to the import of Kepler's descriptive laws. He could scarcely fail to be so, as he attributed activity in the heavens to virtues, powers and souls residing in the heavenly bodies. By a virtue that it emitted the sun swept the planets around, and Kepler allowed them no endeavour to escape its influence. Though the souls of the planets interacted with that of the sun to produce the ellipses (which he felt he must particularly explain) they required no restraint. The moon, the only satellite known in 1609, presented a rather different problem since it did not revolve about the sun, and so (as with Newton in 1666) it was the Earth-moon system that brought Kepler nearest to the idea that the moon has weight with respect to the Earth. But he never quite attained it.

Kepler's speculations attempted to account for the laws of ellipticity he had already discovered ; with the equally important Third Law (1619)* these formed thereafter the necessary conditions of any cosmological hypothesis. The Third Law—whose history from 1619 to 1665 is quite obscure—held the key to the problem of the stability of the solar system. For it indicated in what way the motions of the five planets were mathematically interdependent, so that they formed a harmonious system. It was Newton's genius to use this key, perceiving that interdependence was the product of the action of a single force.

That Newton in 1665 or 1666 was aware of Kepler's Third Law at all is intriguing. Amid the deep silence concerning Kepler's work from 1620 to 1665 a few English astronomers had discussed his theory of the orbit without noticing the Third Law. Possibly Barrow directed Newton to Kepler, for there were copies of all the relevant books in the Trinity Library ; if so, it was crucial advice. Newton did what Descartes had failed to do ; assuming the orbit to be circular, he calculated the force with which the planet would tend to escape from it.† Expressing his measure of rotational force

* That the ratio between the square of the periodic time and the cube of the distance from the sun is the same for each planet $\left(\dfrac{T^2}{r^3}=k\right)$.

† In solving this problem of pure mechanics Newton had been anticipated by Huygens, who published his work seven or eight years later in *Horologium Oscillatorium* (1673), thus again anticipating Newton in priority of publication.

cumbersomely at first,[3] Newton soon found his way to a simpler argument and the formula found in a paper perhaps written about the end of 1666, or even a little after that. Here Newton correctly defined the centrifugal force as that which, applied during the time taken to complete one revolution, would impel any body (such as a planet) from rest to the distance π^2 multiplied by the diameter of the circle (or orbit).[4] From this relation Newton settled Galileo's problem, why bodies are not flung off the spinning Earth ; at most the force of rotation was only about $\frac{1}{350}$th of the force of gravity. Next Newton computed the force of the moon's rotation which, taking the diameter of the Earth to be a little over 6600 miles, proved to be less than $\frac{1}{4000}$th of the force of gravity. And finally a very simple operation showed that from Kepler's Third Law the centrifugal forces acting upon the five planets—the measure of their endeavour to break away from their orbits—were in the inverse ratio of the squares of their distances from the sun.[5]

These were results of tremendous importance. For Newton saw—though he did not explicitly say so—that a single force in the sun, decreasing as the square of the distance from it, would suffice to hold all the planets in their orbits (still assuming that these were circular). Furthermore, it was clear that the force required to hold the moon in its path was rather less than $\frac{1}{60^2}$ of the force of gravity (had he taken 8000 miles as the diameter of the Earth the answer would have been exactly $\frac{1}{60^2}$). From this, since the moon is distant from the Earth sixty Earth-radii, it would also follow that the inverse-square law would apply to the moon, if the force holding the moon in its orbit were the force of gravity familiar at the Earth's surface. Kepler was right, the moon would fall like a stone—if it were not supported by the force of its own revolution.

All this Newton had accomplished by about the end of 1666. He had made an onslaught upon a new branch of mechanics, the dynamics of revolving bodies, and he had at once gone on to consider the planets as bodies moving in a dynamic system and subject to the relations he had just discovered. This was Newton's great step, and was presumably his object from the first. Huygens had gone far in the same direction, but he had not seized upon the idea that sun and planets constitute a dynamic system, subject to mathematical laws (not the merely mechanical laws of Descartes). Precisely because Huygens never outgrew the Cartesian principles of explanation he could not do this ; to him the celestial motions were effected by mechanical, aetherial pressures and therefore they could not be subject to simple mathematical laws like the inverse-square law. Newton's reasoning was founded upon Descartes' law of inertia and Galileo's law of acceleration, but the Cartesian aetherial mechanism adopted by Huygens was expressly designed to prevent the law of inertia's upsetting the planetary orbits, and to prevent their acceleration from the centre. Huygens could never see the problem in Newton's purely mathematical terms, even after the *Principia*. Newton, on the other hand, setting aside mechanistic explanations of gravity, attended only to the mathematical relationships. That stated in Kepler's Third Law declared how things are, and that he derived from pure mechanics showed what the forces must be. They matched to yield the inverse-square law, a law that simple algebra dictated once physical relationships were expressed in mathematical terms. And the calculation for the moon suggested that the centripetal force of the sun holding its system together was the same as the force of terrestrial gravity.

What Newton had done so far was not mathematically complex. He took his measurements from books—the figure for the size of the Earth came from Galileo, and was markedly too small. The dynamical principles he welded into his definition of centrifugal force he took from the common post-Galilean stock. The rest was simple geometry and algebra. But it required brilliance to see how it could be done, and the most powerful scientific insight to see why it should be done. Having come so far, to press on was mathematically far more difficult. For the planets move in ellipses ; they, and the Earth, sun and moon are not the point-bodies of mechanical theory but solid masses. Did they nevertheless behave as if they were points? The agreement in the case of the moon's motion was poor, and the moon's motions were fearfully irregular—astronomers had not learnt to compute them accurately. If Newton committed himself to his new idea of gravitation, he would commit himself to the opinion that somehow the gravitational force must cause the perturbations of the moon's orbit. So far gravity only gave him a neat circular revolution. If Newton thought of Jupiter's satellites, there were similar problems there. And there was the problem of the tides, another roughly periodic motion which no one had reduced to good mathematical order. Worst of all, celestial dynamics did not have to consider two bodies alone—a sun and planet—but a system of a dozen or more, dominated by the sun indeed, yet, if the hypothesis of universal gravitation were correct, all pulling on each other, each disturbing the elegant orbits of the others. With all his might Newton never completely solved this problem so that he could be sure all possible disturbances of the pattern were self-nullifying; he always contemplated the possibility that a miracle might be needed from time to time to prevent cumulative disruption of the heavenly order.

How far Newton looked ahead to these problems in the early Cambridge years he did not say. The fact that he turned from planetary dynamics to pure mathematics and optics suggests that he was aware of some of the difficulties. Possibly he overestimated their complexity, when his own new methods were still imperfect (in fact he was scarcely to use them in writing the *Principia*). In any case he knew he would require far more rigorous mathematical proofs than any he had devised up to this time. In addition he needed more facts; astronomical measurements were still unreliable. Nor could Newton at this stage be sure that Cartesian mechanism was wholly false. One thing he gained from his work in optics was the strengthening of his early anti-Cartesian convictions, and of his belief that the mathematical way in science was the only one leading to truth.

While Newton kept silent others began to catch him up. G. A. Borelli, the mechanistic physiologist of the Accademia del Cimento, published in the very year 1666 the *Theory of the Medicean Planets*, which Newton was to praise later. Clearly the application to Jupiter's satellites was a subterfuge : Borelli was speaking of the whole solar system. Rather in Kepler's manner he imagined that the planets were revolved by the light-rays emanating from the rotating sun. Since this force was weakened by diffusion it grew feebler with distance so that (as Borelli erroneously supposed) the planets would revolve at speeds proportional to the force acting upon them, hence to their distance from the sun. From Descartes Borelli drew the lesson that an inward pull was required to balance the centrifugal force of rotation. He accordingly supposed that each planet had a constant tendency or desire to approach the sun, and to account for the ellipticity of the orbits (for Borelli was the first cosmologist to follow Kepler's Laws) he imagined further that the planet oscillated slowly, like a pendulum, on either side of its mean path. Apart from this last point, the main departure of Borelli's hypothesis from Descartes'—than which it was no more mathematical—was his suggestion of a centripetal force, which he did not dare to call an attraction and did not think of identifying with gravity, that was nevertheless hardly distinguishable from an attraction.

Borelli's universe, like that of Kepler and Descartes in their different ways, was a driven machine, spun round by the sun's vigour. As such it was far harder to analyse mathematically than a system in which the planets were supposed to spin freely. Only if theories of aether-vortices and solar emanations were abandoned would it be possible to return to Galileo's idea of free-spinning planets, with the proviso now known to be required that they must at the same time be subject to an inwards pull. With such a pull (and assuming always the absence of any resistance) it was now clear that the rectilinear, inertial motion of a planet would be bent round into a closed curve. Centripetal force was the factor making " rotary inertia " (as Galileo had conceived it) possible ; that is, the idea of an inwards pull negated the notion of a material aether, and also made such an aether as a celestial driving-force unnecessary.

Robert Hooke saw his way through some of this argument, though not through all for he did not in fact renounce aetherial physics. His views, the product of a lively and often sound imagination in science, developed during the long years of Newton's reticence. He had little more to go on than intuition, the example of Descartes' system, and the analogy with the motion of a conical pendulum. Hooke was no more than an ordinary geometer. He did not even perceive until it was too late, and then with high resentment, that the question he was tackling was one of mathematical, not experimental, physics. He did not know how to calculate centrifugal forces till he learnt it from Huygens, and the intricate mathematics of motion in ellipses was utterly beyond him. Whereas Newton's theory grew out of mathematical relationships, Hooke's grew from qualitative consideration of the effects of gravity.

In the early 1660s Hooke made futile experiments to detect variations in gravity ; in *Micrographia* (1665) he conjectured that the moon might have a gravitating principle like the Earth's ; in the next year, improving upon Borelli, Hooke supposed that a " direct [inertial] motion " might be bent into a curve " by an attractive property of the body placed at the centre " ; and in a tract on the comet of 1677 he wrote :

> I suppose the gravitating power of the Sun in the center of this part of the Heaven in which we are, hath an attractive power upon all the bodies of the Planets, and of the Earth that move about it, and that each of those again have a respect answerable. . . .

What follows makes it clear that in this place Hooke did not identify the attractive power with gravity, although in another tract four years earlier he had done so. There he had proposed to explain

> a System of the World differing in many particulars from any yet known, answering in all things to the common Rules of Mechanical Motions,

which was to be grounded on three suppositions. The first of these was the earliest statement of universal gravitation: that all celestial bodies have an attraction or gravitating power towards their own centres, whereby they attract their own parts and all other bodies within their sphere of activity. Hooke specifically said that sun, moon, Earth and planets have an influence on each others' motions. The second supposition was a statement of the law of inertia, with the special exception that the rectilinear path might be bent into a " Circle, Ellipsis, or some other more compounded Curve Line " by " some other effectual powers " acting on the moving body. The third supposition was that the attractive power diminishes with distance, but Hooke added that he had not experimentally (!) discovered the degrees of this decrease.[6] (Such a remark shows that Hooke's was not the mathematical way.) Though Hooke's hypotheses are very significant in relation to the theory of Newton's *Principia*, there is no evidence that his contemporaries were deeply impressed by their display in lectures before the Royal Society and printed pamphlets, where indeed they were interspersed with many other less prescient hypotheses. He never put together the system of the world that he promised, nor was he ever able to offer any firm support, either experimental or mathematical, for the hypotheses on which it was to be founded. In this Hooke was undoubtedly unlucky, for he was the only man of his age to proceed, broadly speaking, along the track that led Newton to his goal. And it is sad that Newton's irritation at Hooke never allowed him to acknowledge that fact.

Meanwhile Newton himself, who was concerned with other matters and would have found nothing new in Hooke's ideas, took no notice of them. Early in 1673 he had offered his resignation to the Royal Society ; Oldenburg had refused it, and (needlessly) had had Newton relieved of the payment of his subscription, remarking

I could heartily wish, you would pass by the incongruities, y^t may have been committed by one or other of our Body towards you, and consider, that hardly any company will be found in the world, in wch there is not some or other y^t wants discretion. You may be satisfied, that the Body in general esteems and loves you. . . .[7]

Newton's reply was not encouraging. " I intend ", he declared, " to be no further sollicitous about matters of Philosophy ", and he begged to be excused from the writing of further replies to " philosophical letters ". Nevertheless he did in the next two years answer renewed objections to his optical theory, and submit the further long paper on the colours of thin plates to the Royal Society. After that, which had provoked a fresh claim by Hooke that the subject had all been covered in *Micrographia* and a furious rejoinder from Newton, he had little further connection with the Society in the years after 1676. Oldenburg, Collins and Barrow were all soon dead, and though Newton's friendship with Boyle became warmer, the overtures from London ceased. Hooke and Nehemiah Grew became the Society's secretaries, and the *Philosophical Transactions* lapsed for a while.*

After their latest conflict, Hooke had written a conciliatory letter to Newton in January 1676, laying the blame on Oldenburg's misrepresentations, professing admiration for Newton, and declaring himself " well pleased to see those notions promoted and improved which I long since began, but had not time to compleat ". Without servility or surrender Hooke offered private correspondence in place of public dispute. Newton replied with equal civility and no greater sincerity, even assuring Hooke :

you defer too much to my ability for searching into this subject [the theory of colours]. What Des-Cartes did was a good step. You have added much several ways, & especially in taking

* Grew continued the *Transactions* for a few months. After the lapse Hooke published seven issues of *Philosophical Collections* between 1679 and 1682. In the next year the *Transactions* were revived by Robert Plot, the new secretary.

yᵉ colours of thin plates into philosophical consideration. If I have seen further it is by standing on yᵉ shoulders of Giants.[8]

(In the fable the keen-sighted pygmy saw further by being raised on a giant's shoulders ; Newton was ingeniously praising and deprecating his achievements at the same time.)[9] But the olive branch did not sprout. The two men were too close in their interests and too alien in their understanding of what science is for amity to be possible ; both were unduly touchy, intellectually arrogant and sensitive to their rights of priority. Newton was in general honest (or cautious) in acknowledging experiments or measurements borrowed from others ; he was habitually careless in appeasing the vanity of lesser men whose ideas, momentous to their progenitors, merely paralleled early stages of his own thought or furnished seeds for his own grander conceptions. And it was Hooke's misfortune that he was not satisfied with being what he was—which was very considerable—but wanted to be Newton too, which he was not.

The two men did not quarrel. A few notes on business passed between them after Hooke succeeded Oldenburg as Secretary of the Royal Society. Two years later (November 1679) Hooke again prompted Newton to warmer relations. He wrote the chatty kind of letter a secretary might write to an absent and discontented member of his society. It had only one important sentence, in which Hooke asked Newton's opinion of his own hypothesis " of compounding the celestial motions of the planets of a direct motion by the tangent and an attractive motion towards the central body ". Abruptly, Newton's attention was drawn to mechanics, and to the recollection of scattered papers neglected for twelve years. Though in his reply Newton disclaimed awareness of Hooke's hypothesis, and avowed that he had for some years past turned his mind from science to other studies " in so much that I have long grutched the time spent in that study unless it be perhaps at idle hours sometimes for a diversion ", he could not conceal the fact that his attention had been caught.* For he went on to suggest a "fancy of my own" about the fall of a heavy body to the centre of the Earth, supposing there were space for it to fall the whole way.

Now there was a spate of letters, from which the *Principia* was born. Sacrificing tact to his desire to score a point Hooke rejected Newton's spiral trajectory, terminating at the centre of the Earth, as a vulgar error. Unless resisted by air the falling body would return to its point of origin, describing "a kind of Elleptueid" (Hooke did not know that the curve would be an exact ellipse). Not content with that, he read the letters exposing Newton's mistake to the Royal Society. Newton was angered; eager to prove Hooke wrong, he computed the trajectory of descent *assuming that the force of gravity was uniform* between the surface and the centre ; the result was a sort of clover-leaf curve, nothing like an "elliptoid". Once more Hooke caught him out. What Newton wrote was correct, he retorted (it was now 6 January 1680) :

But my supposition is that the attraction [to the centre] always is in a duplicate proportion to the Distance from the center reciprocall [i.e. as $\frac{1}{d^2}$], and Consequently that the Velocity will be in a subduplicate proportion to the Attraction, and consequently as Kepler supposes Reciprocall to the Distance. . . .[10]

Hooke went on to say that while the question under debate was trivial in itself, when applied to the celestial bodies this "curve truly calculated will shew the error of those many lame shifts made use of by astronomers to approach the true motions of the planets with their tables". From the dynamical conditions he had defined—peripheral motion and central attraction obeying the inverse-square law—it should be possible to calculate the true path of a planet which would prove to be, Hooke thought, like but not identical with the

---

* Was Newton deliberately misleading or not ? Earlier in the year he had written the famous *Letter to Boyle*, he had been busy with chemistry, and he *had* read Hooke's tracts. But the years 1676-9 are ones of which little is known in Newton's life.

regular ellipses assumed by astronomers. (If he had supposed that Kepler's ellipses were true orbits, Hooke would obviously have used the word ellipse before and would have had no doubt as to the nature of the curve he sought.) Newton just ignored this letter. Eleven days later Hooke, hardly aware of the mortal offence he had given, appealed to him again, in terms that reveal serious inadequacies in his concept, to discover the curve :

> I doubt not that by your excellent method you will easily find out what this curve must be, and its proprietys. and suggest a physical reason of this proportion.[11]

But Newton was not going to be another man's computer. He kept his silence, and what Hooke had made him discover remained hidden for another four and a half years.

Hooke was never able to find the demonstration that Newton denied him. Given his concept of universal gravitation, when Huygens' theorems on centrifugal force were printed in 1673 he had the wit to combine them with Kepler's Third Law to deduce the inverse-square law of attraction, as Newton had done long before in 1666, *for circular motion*. He guessed that in the heavens this law would 'cause the ellipse-like motion required by Kepler's First and Second Laws, but—regarding the matter physically and not mathematically—he assumed that the curve would be more complex than a geometrical ellipse, and so he told Newton that the curve of the body falling towards the centre of the Earth be an " Elliptueid ". (In this Hooke curiously paralleled the development of Kepler's own ideas, for Kepler had at first thought that the orbit would be more complex than a perfect ellipse.) Hooke was no Platonist. He challenged Newton to define this more complex curve ; what Newton actually found was a geometrical demonstration of motion in an ellipse. Kepler's First and Second Laws were, dynamically, perfectly exact, not the approximations Hooke had taken them to be. And of this he left Hooke ignorant.

How the *Principia* came to be written is a well-known story. By 1684, besides Newton and Hooke, Halley and Wren were intuitively convinced of the inverse-square law of gravitation. None of the three last could derive orbits from that law mathematically. Halley, a young and energetic man, made a special journey to Cambridge to put the problem to the mathematical professor : " What would be the curve described by the planets on the supposition that gravity diminished as the square of the distance ? "

> Newton immediately answered, *an Ellipse*. Struck with joy and amazement, Halley asked him how he knew it ? Why, replied he, I have calculated it ; and being asked for the calculation, he could not find it, but promised to send it to him [Halley].[12]

Newton had carelessly mislaid what other brilliant men with their best efforts could not find out ! Re-working his calculation, Newton realised it could hardly stand by itself ; it required, to be convincing and clear, axioms, definitions, subsidiary propositions, in short a demonstration of the use of mathematical reasoning in handling problems of mechanics. Within a few days or weeks he had decided to devote his lectures of the approaching Michaelmas Term, 1684, to *The Motions of Bodies*. Two or three months later, in October, he began to read a text that is substantially that of Book I of the *Principia*. Next month some of its propositions were despatched to Halley, in fulfilment of his promise, who hastened off to Cambridge a second time to persuade Newton to lay his work before the Royal Society. His wish prevailed, and Newton set to work to revise and expand his lectures, but the *Philosophiae Naturalis Principia Mathematica* (" Mathematical Principles of Natural Philosophy ") assumed its ultimate form only gradually during the next two years.

In 1686, when Book I was duly presented to the Royal Society, Hooke and Newton quarrelled for the fourth and final time. It was a quarrel that led to Hooke's virtual retirement from the Society during the last years of his life ; that strengthened Newton's desire to abandon the university and science ; and that prevented the publication of *Opticks* until after Hooke's death. No doubt Hooke's charge of plagiarism against Newton was no more unjust than Newton's neglect of Hooke's just claims to recognition. Yet Newton could rightly assert that he had far surpassed Hooke—no comparison between their achievements is possible ; his counter-charge that he had learned nothing from Hooke, who had proved nothing, was fair ; but his jealous temperament swayed him into injustice when he proceeded to deny that Hooke had accomplished anything at all. In the course of his outpourings against Hooke, in letters to Halley (who had undertaken to pay for the printing of the *Principia* and see it through the press), Newton made some remarks of more than partisan importance underlining the obstacles that he had surmounted before the *Principia* could be written. The first, naturally, was the proof that the inverse-square law yielded an elliptical orbit.* To the discovery of the law itself he attached little importance, since it was obvious after Huygens' book of 1673. The difficulty was to prove it for the astronomical ellipses as a mathematically exact law, not to infer it for circles.

> There is so strong an objection against the accurateness of this [inverse-square] proportion [Newton added] that without my demonstrations, to which Mr Hooke is yet a stranger, it cannot be believed by a judicious philosopher to be anywhere accurate.[13]

By this Newton meant that on the evidence any competent physicist would recognise the impossibility of the law's exactness ; only a mathematician could turn the tables and demonstrate that —paradoxically—it was perfectly exact and physically sufficient. Kepler, said Newton (rather unreasonably), had guessed that the planetary ovals were geometric ellipses ; his own demonstrations proved at one and the same time the mathematical precision of both the inverse-square law and the ellipses.

Secondly, Newton recollected,

I never extended the duplicate proportion lower than to the

---

* Or more generally, a conic which may be closed (ellipse and circle) or open (parabola).

superficies of the Earth, and before a certain demonstration I found the last year [1685], have suspected it did not reach accurately enough so low.[14]

It was a ready inference for Newton that the gravitational force must reside in the ultimate particles of matter, if it is a property of gross bodies. Now, even such astronomically adjacent bodies as the Earth and moon are so distant that the lines drawn between any two particles in these bodies could be considered parallel and of the same length, hence it was easy to see that the total force was the sum of the forces of the particles. In the case of a body only a few feet above the Earth's surface, like an apple on a tree, the situation was quite different. How could the summation of particulate forces be effected in this case, and where would the centre of force in the Earth be located ? In the *Principia* Newton confessed his doubt that the commonsense way of thinking would work :

> After I had found that the force of gravity towards a whole planet did arise from and was compounded of the forces of gravity towards all its parts, and towards every one part was in the inverse proportion of the squares of the distances from the part, I was yet in doubt whether that proportion as the square of the distance did accurately hold, or but nearly so, in the total force compounded of so many partial ones ; for it might be that the proportion that was accurate enough at greater distances would be wide of the truth near the surface of the planet, where the distances of the particles are unequal, and their situations dissimilar.[15]

Doubt was removed by a group of theorems in which Newton integrated mathematically the individual forces arising from the infinitely numerous particles of two solid spheres.* They proved that the centripetal (or gravitational) force between two spheres " increases or decreases in proportion to the distance between their centres according to the same law as applies to the particles themselves. And this is a noteworthy fact." [16] Once more mathematics

* *Principia*, Book I, Propositions LXX to LXXVII.

demonstrated a physical improbability as truth : the sphere acted on any body outside it, however close, from its own centre ; and this was true of any law of attraction. But if the inverse-square law applied outside the sphere, then inside it the attraction to the centre was directly as the distance.

Here were at least two instances where the physical or imaginative implausibility of the inverse-square law could only be corrected by mathematical reasoning. Newton was claiming, in effect—and he was right—that his great achievement lay not in imagining a physical hypothesis (to which there were grave objections that qualitative physical thinking could not surmount), but in proving mathematically that what seemed to be objections were on the contrary, when properly analysed, decisive testimonies to the accuracy of the theory. In other words, the theory of universal gravitation was only worth anything when it was a mathematical theory ; and only as a mathematical theory could it be verified by observation in such a way as to sway conviction. Hooke, from the opposite pole, made a remark that provoked one of Newton's most furious outbursts :

[Hooke] has done nothing, and yet written in such a way, as if he knew and had sufficiently hinted all but what remained to be determined by the drudgery of calculations and observations, excusing himself from that labour by reason of his other business, whereas he should rather have excused himself by reason of his inability. For 'tis plain, by his words, that he knew not how to go about it. Now is not this very fine ? Mathematicians, that find out, settle, and do all the business, must content themselves with being nothing but dry calculators and drudges ; and another, that does nothing but pretend and grasp at all things, must carry away all the invention, as well as those that were to follow him, as of those that went before.[17]

Imagination, the ability to feign hypotheses, could give only the beginning of a theory in physics. To find out, to do everything, was to make a mathematical theory and confirm it by experiments or observations.

That is why Newton entitled his great work *The Mathematical Principles of Natural Philosophy* rather than, as he once thought, simply *On the Motion of Bodies*. To Newton, the laws of nature were not certainties of introspection, but those derived by mathematical reasoning. The method of explaining phenomena by reference to these laws was not by ingenious hypotheses, but again by mathematical reasoning. Galileo's proclamation of faith, tinged still with numerical or geometrical mysticism of Pythagorean or Platonic origin, became for Newton a plain, stern rule of procedure :

the whole burden of philosophy seems to consist of this : from the phenomena of motion to investigate the forces of nature, and then from these forces to demonstrate the other phenomena.[18]

To follow any other course than the mathematical in this was quite simply to fail to comprehend the nature of physics. Yet even Newton was not quite free of the ancient delusion that mathematics is something more than logic, that it has in itself the roots of harmony and order, since he could find repeated in the colours of the spectrum the ancient divisions of the musical chord.[19]

The *Principia* was, and is, a difficult book. Few of Newton's contemporaries were capable of working systematically through it, so that his celestial mechanics became widely known either through popularisations like Henry Pemberton's *View of Sir Isaac Newton's Philosophy* (1728) or through the " translation " of its theorems into the language of the calculus. As Newton wrote at the opening of Book III, *The System of the World*,

. . . not that I would advise anyone to the previous study of every Proposition of [the First and Second] Books ; for they abound with such as might cost too much time, even to readers of good mathematical learning. It is enough if one carefully reads the Definitions, the Laws of Motion, and the first three

Sections of the First Book. He may then pass on to this
Book. . . .[20]

John Locke, it is said, sought an assurance from Huygens that
Newton's mathematics could be relied upon ; he then read the
Principia for its scientific ideas (which he adopted). No doubt many
others approached it in the same way. Within only a few more
years another difficulty came between Newton and the reader.
Newton had the ill-luck to write at the moment when the new
mathematics was born, at the very moment of Leibniz's first paper
on the calculus. Within a generation continental mathematicians
had adopted Leibnizian methods of handling such problems as
Newton had treated in the *Principia*, while the British were turning
rather less rapidly to Newton's own fluxional analysis (p. 95).
But Newton's mechanics was fossilised in the geometrical proofs of
an older generation of mathematicians—Huygens, Wallis, Barrow,
Gregory ; in a sense it was the last great piece of mathematical
physics composed in the Greek tradition. Its form was soon to
seem both old-fashioned and laborious.

Nor were the difficulties merely mathematical. To mathematicians comparable in stature to Newton, such as Huygens and
Leibniz, it was his philosophy that proved unacceptable. Did the
theory of gravitation assume that bodies are capable of acting upon
each other at a distance without mechanical connection ? Were
space and time to be conceived of as Newton required ? Was he
perverting the apparently triumphant metaphysic of mechanism ?
Did he make God, as Leibniz alleged, an imperfect workman who
needed to tinker continually with His creation ? These were questions that transcended mathematical reasoning and experimental
decision, yet until they were settled the status of Newton's work
remained, in a measure, in suspense.

The *Principia* opens with twenty-five pages of *Definitions* and
Axioms stating the basic concepts of mechanics and subsuming under
the most general principles virtually all that had been accomplished

in that science before 1687. Here Newton for the first time gives dynamics a clear, coherent foundation such as neither Galileo nor Huygens had offered. He defines mass, momentum, inertia, force and centripetal force, remarking of the last that he uses the words *attraction*, *impulse* and *propensity* indifferently since he considers centripetal forces not physically but mathematically. These definitions are followed by the famous scholium on space and time further discussed below. The axioms consist of three laws of motion and six corollaries. Law I is the law of inertia ; Law II states that accelerations is proportional to force ; Law III is the principle of the equivalence of action and reaction.* The first two corollaries explain the parallelogram of forces and its application ; the third and fourth assert that the total momentum and the centre of gravity of a system of bodies are unaffected by their mutual actions, while the fifth and sixth state that such actions are in turn unaffected by uniform or uniformly accelerated motion. In a second scholium Newton next gives examples of the use of the axioms in obtaining specific results discovered by his predecessors in mechanics ; after having thus concisely laid the groundwork he proceeds to his own advance beyond them.

Book I of the *Principia* provides the complete mathematical theory on which celestial mechanics rests. It is wholly general, and in the earlier propositions a geometrical mass-point takes the place of a physical body ; here Newton analyses the relations between orbits and central forces of different kinds. The outstanding result is the proof that the orbit is a conic, with the centre of attraction at one focus, if the force varies with the inverse square of the distance. The conditions of motion in such orbits are precisely laid down,

---

* " I. Every body continues in its state of rest or of uniform rectilinear motion unless compelled to change its state by the action of forces.

II. The change of motion is proportional to the force acting, and takes place along the straight line along which the force acts.

III. There is always a reaction equal and opposite to action ; or, the actions of two bodies on each other are always equal and opposite."

and Newton shows how to determine the curvature of an orbit from a few observations of position. This was a practical problem for astronomers, as was the problem of the motion of the axes of the orbit, which he also treats. He gives also an approximate solution for finding the motions of three mutually attracting bodies, such as the sun, Earth and moon. Then he deals with the attractive forces of spherical and aspherical bodies arising from their component particles, from which it follows that the heavenly bodies can be reduced (externally) to point-masses as treated in the earlier theorems. Book I also contains a section on the motion of pendulums (relating to the theory of the Earth's shape) and closes with a discussion of the motion of particles when attracted by large bodies, from which Newton draws conclusions he thought applicable to optics.*

The first Book was completed in the spring of 1686, and sent to the Royal Society : Newton had also worked on Books II and III but these were as yet in nothing like their final form. At first he intended to write Book III—containing the application to astronomy of the relationships established in the first Book—" in a popular method, that it might be read by many " but the quarrel with Hooke induced him to reconstruct it in the mathematical way.†
Meanwhile Book II, short at first, was considerably expanded by extra theorems. It is an interpolation in the main argument, dealing with the motions of fluids, and of bodies in fluids, in fact a treatise laying the foundations of fluid mechanics. Here the mathematics of the *Principia* reached its highest complexity, and indeed in the first

---

* Newton was careful to deny that he identified light-rays with streams of moving particles ; he was not " inquiring whether they are bodies or not ; but only determining the curves of bodies which are extremely like the curves of the rays." [21] However, these propositions give Newton's best account of the mechanism of refraction-reflection-inflection, and he seems to have taken them as physically valid.

† It is practically certain that the separate work *On the System of the World*, posthumously printed, is the original Book III. It too was given as a course of lectures at Cambridge. Newton never lectured, apparently, on the content of Book II.

edition Newton fell into a series of notable errors of mathematical reasoning. Book II also contains experimental proofs of many results (like the speed of sound) that Newton derived mathematically from purely theoretical considerations, basing himself throughout on the mechanical philosophy. Near the end he demonstrated conclusively that Cartesian vortices could not account for the observed planetary motions. Book III, continuing the astronomical theme, shows that from the masses, distances, and velocities of the sun, planets and satellites the theory of Book I predicts all the known phenomena, if gravity is postulated as the universal centripetal force obeying the inverse-square law. Newton took particular trouble with the involved motions of the moon (of which however he did not quite complete the theory), and with the theory of the tides. He established the orbits of comets as either parabolas or highly eccentric ellipses (figure 14), and computed the shortening of the Earth's polar axis caused by its axial rotation.

No other work in the whole history of science equals the *Principia* either in originality and power of thought, or in the majesty of its achievement. No other so transformed the structure of science, for the *Principia* had no precursor in its revelation of the depth of exact comprehension that was accessible through mathematical physics. No other approached its authority in vindicating the mechanistic view of nature, which has been so far extended and emulated in all other parts of science. There could be only one moment at which experiment and observation, the mechanical philosophy, and advanced mathematical methods could be brought together to yield a system of thought at once tightly consistent in itself and verifiable by every available empirical test. Order could be brought to celestial physics only once, and it was Newton who brought order. His is the world of law. Since everything that happens in this world is the effect of motion, the primary, never-failing laws are the laws of motion defined at the beginning of Book I. Motion—except in the rare event of pure inertial motion—is the product of force ; therefore in physics the next set of laws should

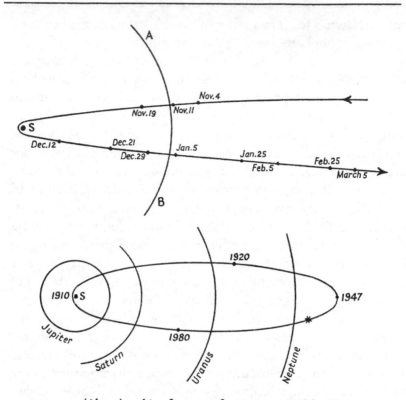

FIG. 14. (Above), orbit of comet of 1680, computed by Newton in the *Principia* (s=sun, AB=Earth's orbit); (below), present orbit of comet of 1682 (Halley's Comet)

define the forces that operate in nature. Of these Newton succeeded in defining only one, the law of gravity: the *Principia* is for the most part his treatise on this one force and the phenomena that arise from it. In it, such older descriptive laws as Kepler's and Boyle's are deduced as consequences of the basic law of force (though in the latter case the force is not that of gravity).* Thirdly, there are the

* Newton's deduction of Boyle's Law from a repulsive force between the particles of a gas varying inversely as the distance (Book II, Proposition XXIII) must be considered a noble failure, however.

laws of mathematics, belonging to the sphere of reason and logic indeed rather than to the sphere of physical reality, yet laws which physical reality obeys, and which scientific reasoning must not ignore. Nothing happens by chance, nothing is arbitrary, nothing is *sui generis* or a law unto itself. The philosophy of both *Principia* and *Opticks* insists that however varied, disconnected and specific the almost infinite range of events in nature may seem to be, it is so in appearance only : for in reality all the phenomena of things and all their properties must be traceable to a small set of fundamental laws of nature, and by mathematical reasoning each of them is deducible again from these laws, once they are known.

Yet Newton shrank from the belief that these laws are innate in nature ; that in his view would lead to necessitarianism and the deification of matter. Matter and material properties could not be eternal and uncreated ; rather matter is, and the laws of nature are, because God has willed them. The perfection of the laws implied for him a lawgiver, as the perfection of the architecture of the universe implied a cosmic design :

> though [the planets and comets] may indeed continue in their orbits by the mere force of gravity yet they could by no means have at first derived the regular positions of their orbits from those laws . . . it is not to be conceived that mere mechanical causes could give birth to so many regular motions, since the comets range over all parts of the heavens in very eccentric orbits . . . and in their aphelions, where they move the slowest and are detained the longest, they recede to the greatest distances from each other, and hence suffer the least disturbance from their mutual attractions. This most beautiful system of the sun, planets, and comets could only proceed from the counsel and dominion of an intelligent and powerful Being. And if the fixed stars are centres of other like systems these, being formed by the like wise counsel, must be all subject to the dominion of One ; especially since the light of the fixed stars is of the same nature as the light of the sun, and from every

system light passes into all the other systems ; and lest the systems of the fixed stars should, by their gravity, fall on each other he hath placed those systems at immense distances from each other.[22]

The ideas that Newton expressed in the General Scholium with which he concluded the second edition of the *Principia* (1713) had persisted throughout his life, for they occur in a document he wrote as a young man, before 1669. Like Descartes and Boyle, Newton saw the mechanistic universe as an argument against atheism, not in favour of it. But Newton's God—it will now be clear—was no Christian deity, nor the God of any sect. The attributes Newton conferred upon him are unexceptionable : he is living, intelligent, omnipotent, eternal, omniscient and most perfect, but he is a God of science, not theology. His kingdom is in the brain rather than the heart, for he is a God of law and certainty, not a God of hope and fear, of punishment and reward.

Even in a universe of law men can know only what is relative and superficial. They cannot detect the immutable, unvarying flow of duration, wrote Newton, or the unchanging extension of space. They can know time only by observing the succession of events, and space by measuring the distances between bodies without knowing if one or many among them move. Neither can men know the " inward substances " of material things. Only God is absolute and perceives the absolute. True, " He is not eternity and infinity, but eternal and infinite ; he is not duration and space, but he endures and is present ".[23] Nevertheless, because he " exists always and everywhere " God is for Newton the guarantor of the absoluteness of time and space, for absolute time and absolute space are the dimensions of God as their relative counterparts are man's. In a long Scholium following the Definitions of matter, motion and force with which the *Principia* begins Newton sought to clarify the distinction between absolute and relative dimensions, as he thought was necessary to avoid " prejudice ". Indeed the distinction is for Newton a vital one, for the absolute are to him the

YXI was to PX⁹ as SU⁹ (that is (by ~~the~~ Cor. Lem. 2) AB × PQ to
GHI, whence invertedly YXI is to AB × PQ as PX⁹ to GHI & by
consequence as YZ⁹ to KLI. w⁰ w. to be Dm.   3

Prop. III.

If a body be attracted towards either focus of any Ellipsis & by that attraction be made to revolve in the Perimeter
of y⁰ Ellipsis: the attraction shall be reciprocally as the square
of the distance of the body from that focus of the Ellipsis.

Let P be the place of the body ~~along~~ in the Ellipsis
at any moment of time & PX the tangent in w⁰⁰ the body
would move uniformly were it not attracted & X y⁰ place
in that tangent at w⁰⁰ it would arrive in any given part
of time & Y the place in the perimeter of the Ellipsis
at w⁰⁰ the body doth arrive in the same time By means of
the attraction. Let us suppose the time to be divided into
equal parts & that those parts are very little ones so y⁰
they may be considered as physical moments & y⁰ y⁰ attraction
or acts not continually but by intervalls ~~acts~~ once in the beginning of every physical moment & let y⁰ first action of
upon y⁰ body in P the next upon it in Y & so on perpetually, so y⁰ PY ~~body~~ may move from P to Y in the chord
of y⁰ arch PY & from Y to its next place in y⁰ Ellipsis in the chord of y⁰ next arch & so on for ever. And
because the attraction in P is made towards F & diverts
the body from y⁰ tangent PX into y⁰ chord PY so that
in the end of the first physical moment it be not found
in the place X where it would have been without y⁰ attraction but in Y being by y⁰ force of y⁰ attraction in P
translated from X to Y: the line XY generated by the
force of y⁰ attraction in P must be proportional to that
force & parallel to its direction that is parallel to PF
Produce XY & PF till they cut the Ellipsis
in I & 2. Joyn FY & upon FP let fall
the perpendicular YZ & let AB be the
long Axis & KL y⁰ short Axis of y⁰
Ellipsis And by the third Lemma YXI
will be to AB × PQ as YZ⁹ to KL⁹
& by consequence YX will be equall to

$$\frac{AB \times PQ \times YZ^9}{XI \times KL^9}$$

And in like manner if py be the chord of another Arch
py w⁰⁰ the revolving body describes in a physical moment of time
& px be the tangent of the Ellipsis at p & xy the subtense of
the

Newton's demonstration, written in English for John Locke, the
philosopher, that the inverse-square law of gravitation holds in an

Detail from the ceiling of the Painted Hall in Greenwich Hospital commemorating the triumphs of William and Mary, the work of Sir James Thornhill (1675-1734). Each end of the ceiling illustrates the arts and sciences related to navigation. Here Copernicus is shown with a mathematician (Galileo?) on his right and some figures from Newton's *Principia*

only truly valid dimensions, while the relative vary with circumstance.

Hence the relative dimensions are not the same as absolute time and space, though they bear the same names, but rather refer to the direct measures of dimension (whether true or false) which we ordinarily use in place of the absolute measurements. If the meaning of words is to be taken from customary usage, then by the terms time, space, place and motion, we should understand direct measures of these quantities, and an expression will be unusual, and exclusively mathematical, if measurements in absolute terms are meant. For this reason, they are violating the sanctity of language who take these words time, place and so on to signify absolute dimensions. And they are equally guilty of defiling mathematical and philosophical truth who confuse absolute measures with their relative equivalents and with direct, physical measurements.[24]

The culprits are, of course, the Cartesians who define motion as displacement from neighbouring bodies. As Newton pointed out, this meant that the particles on the surface of a body moved while those in the interior of the same body were at rest. And their error had more serious consequences.

Absolute time, Newton thought, could be ascertained mathematically by correcting the celestial motions, or perhaps those of some more perfect timekeeper. Absolute space was beyond experience, since no body could be known to be absolutely at rest. Accordingly absolute motion—motion in absolute dimensions— could not be measured directly either. "Yet the thing is not altogether desperate", Newton thought, for he believed that absolute *rotation* was always manifested by its accompanying centrifugal force. The application is obvious, and was to be brought out in the rest of the *Principia* ; now the worst of the Cartesian folly in confusing absolute with relative motion becomes plain. For in the solar system there is a centrifugal force in the planets—otherwise they would fall into the sun under gravity.[25]

Hence the philosopher who observes the due distinction between the species of motion will discern that the planetary motions are absolute, not relative, with respect to the sun and fixed stars. (This is not to deny that relative components of motion are involved also.) At last, by mathematical reasoning upon mechanical principles, the Copernican question was settled : the Earth and planets did revolve, not the sun and stars. This decision was the fruit of the distinction between absolute and relative dimensions, as Newton was to explain more at large in subsequent pages of the *Principia*. " For to this end I composed it."

Newton had solved the greatest scientific problem of his own and the preceding age. In a law-bound universe only one celestial geometry could work, that of Copernicus, Galileo, Kepler and Newton himself. Only God might know whether the sun and stars moved in the mystery of the absolute, but man had proved that the Earth did.

CHAPTER XI

THE AGE OF NEWTON

Gravity must be caused by an Agent acting constantly according to certain Laws ; but whether this Agent be material or immaterial, I have left to the Consideration of my Readers.[1]

The *Principia* made Isaac Newton the greatest figure in European science. Even those who criticised him, with the possible exception of Leibniz, recognised that his stature was greater than theirs. He received the adulation of the Royal Society, and was consulted by philosophers and theologians as the sole authoritative spokesman of scientific thought. The portrait by Kneller, painted about this time, shows Newton with a thin eager face, an introspective look, strong chin, and brows slightly knit ; he wore his own long unruly hair, open shirt and loose gown. It is the picture of an intellectual and is the only portrait of Newton that remotely suggests the author of the *Principia*.[2] He was forty-five and at the height of his powers. Yet with his greatest work Newton bade farewell to science and mathematics. His literary production in the years after 1687, far larger than in the first half of his life, involved no fresh creative effort ; *Opticks* in 1704 and other books published later had been composed long before. Almost as soon as the *Principia* was complete Newton engaged himself warmly in the resistance to the Catholic policies of James II (whose name appears prominently in the book's dedication to the Royal Society). He was elected to the Convention Parliament of 1689 and, as a supporter of the Glorious Revolution, became acquainted with some of the chief Whig politicians. To them he now looked for preferment. An attempt, pursued with all

Newton's customary energy, to obtain for him the Provostship of King's College failed because he was clearly ineligible under the Statutes. He whiled away a few more years at Trinity, writing much on theology and collecting emendations to the *Principia*, until appointed to the Royal Mint in 1696. It was Newton's service to the Crown, not to science, that earned him a knighthood in 1705. He had then been for two years President of the Royal Society, an office he was to hold until his death in 1727 ; he was the first major scientist to be so elected since Viscount Brouncker (President, 1662–77).[3] During a quarter of a century while he dominated the scene almost every scientific activity in England was remodelled upon Newtonian principles and, less happily, was involved in the bitter dispute that had broken out between Newton and the continental mathematicians over the invention of the calculus.

It was a fruitless, mischievous quarrel on which posterity has pronounced in favour of Leibniz and against Newton, whose middle-aged harshness contrasts unpleasantly with the friendliness and admiration of his early letters to Leibniz. The charge that Leibniz had stolen the ideas of the calculus from Newton's letters to Collins and Oldenburg in the 1670s was first made by old John Wallis, ever tender concerning English rights of priority, in 1695 ; it was reinforced in 1699 ; public rancour was sharp about 1705 and again in the years after 1713. The only significance of the affair is that the prejudice against Newtonian mathematics on the continent extended to Newtonian physics, and that Britain was divorced by its partisanship from the main continental stream of mathematics and theoretical physics for over a century.

With some plausibility it has been suggested that Newton's character deteriorated after a severe mental illness of 1693 ; that after his recovery his behaviour was more arrogant, tyrannical and insensitive to other men than it was before. Paranoiac symptoms appear in Newton's letters as early as 1672 ; in 1693 they became temporarily acute, when he suffered from extreme insomnia accompanied by delusions that Pepys was trying to patronise him and Locke to " embroil him with women ". His friends mitigated astonishment with human understanding and in a few weeks Newton came to his senses. The breakdown has been attributed to overwork, but its origins seem rather to lie in a double frustration : the frustration inherent in his unnatural chastity, and of his disappointed hopes. At the age of 51 Newton knew that his work in science was finished ; but his desire for a place in the great world, for renown and power, seemed to be obstructed at every turn. It would have destroyed his hopes altogether to have protested childishly against his fate—as he had vented his feelings at the affronts of 1672, 1676 and 1686, and as he was to do in later years ; the emotions he suppressed broke out in accusations born of delusion against the very friends who were closest to political influence. But it seems unlikely that Newton's character was any more permanently affected than his intellectual vigour, for the personal friends of later years present a uniformly pleasant picture of his affability. If the bleaker side of Newton became more obvious—it was always latent—it may well have been, as so often, because he now had the power to be as disagreeable as he pleased. If his world was a small one, he was its acknowledged sovereign.

His subjects were nearly all Englishmen. Abroad the *Principia* was received at first with cold admiration ; it did not win conviction. Almost the only whole-hearted foreign convert was the eccentric young Swiss mathematician, Fatio de Duillier (1664-1753), who rushed across Europe to place himself at Newton's side. Christiaan Huygens, now in his sixties, disputing amicably with Newton over a number of technical matters, showed not the least inclination to give up his own aetherial theory of gravity, basically Cartesian.* In 1690 Huygens published a *Discours sur la Cause de la Pesanteur* in which he supposed that the revolution of a closed vortex of fluid matter about the Earth, at a speed 17 times as great as that of a point on the Earth's Equator, would press heavy bodies downwards with

---

* A note by him reads : "Vortices destroyed by Newton. Spherically moving vortices instead ".

the acceleration of gravity. The mechanical effect to which Huygens appealed is a genuine one, but he did not explain how this high-speed vortex, though able to press bodies downwards by reaction, failed to carry them along with it ; nor how the vortex could rotate in all directions at once so as to impel bodies in a direction normal to the surface of the Earth. The *Principia*, he wrote, had won " great esteem, and rightly so, for one could not hope to see any more scientific treatment of these questions, nor one that reveals greater intellectual penetration ".[4] He accepted the " hypotheses " (as he called them) of universal gravitation and the inverse-square law and did not doubt that Newton's celestial mechanics, in so far as it was founded upon these, was true. Specifically, it surmounted some obvious objections to the vortex theory of Descartes. On the other hand Huygens could not accommodate himself to the notion of empty cosmic spaces, nor to any suggestion

that gravity may be an inherent quality of corporeal matter. But I do not think that Mr Newton would consent to it either, for such an hypothesis would take us far away from mathematical and mechanical principles.[5]

It remained, then, as the sole possibility that gravity be caused by an aetherial motion such as that he had himself proposed. Huygens' position in fact reduced the *Principia* to a set of mathematical hypotheses, the status Catholic theologians had once assigned to Copernicus' *De Revolutionibus* ; the inverse-square law was for Huygens an odd quantitative relationship that left the central problem of the cause of gravity untouched. Just as a quarter of a century earlier, the point of Newton's work escaped him, for he still did not perceive that mathematical theory was the core of physical explanations. He still demanded mechanisms, cog-wheels, billiard-balls.

Other continental scientists clung equally tenaciously to the pure Cartesian doctrine of mechanism, clung to it long after they knew that the vortices were mere figments like the Ptolemaic orbs.

Indeed, the most popular of all expositions of Cartesian cosmology, Fontenelle's *Conversations on the Plurality of Worlds* (1686), was printed only a year before the *Principia*. Its author, secretary of the Académie des Sciences from 1697 to 1740, was no great brain ; his celebrated squib, running through eighteen editions before 1720, borrowed heavily from John Wilkins' fantastic *Discovery of a New World, or Discourse on the World in the Moon* (1638).[6] This pretty poor stuff was taken for science by the average reader, and Fontenelle himself hardly seems to have been aware that Newton had presented a challenge to his whimsies until after the Englishman was elected to the French academy in 1699.* Thereafter he opposed the ideas of the *Principia* with all his might, and suppressed the work of those within the Académie who, like Pierre Varignon (1654–1722), sought to explore Newtonianism. Many years later, after the death of Newton, Fontenelle in his *Éloge* (the first public account of Newton's life and discoveries) could praise his genius " which in the whole compass of the happiest age was shared only amongst three or four men picked out from all the most learned Nations ", and treat his destruction of the Cartesian vortices with tact. At the same time he thought it would have been better if Newton had spoken of the gravitational impulse rather than the gravitational attraction, and could not resist the jibe that

> Attraction and Vacuum banished from Physicks by Des Cartes, and in all appearance for ever, are now brought back again by Sir Isaac Newton, armed with a power entirely new, of which they were thought incapable, and only perhaps a little disguised.[7]

Fontenelle was not to be moved. He published his final defence of the Cartesian vortices in 1752, at the age of ninety-five.

Nearly half a century had passed since its first appearance before the *Principia* found a popular champion in France, not in a

---

* Newton was the eighth foreign associate to be elected after reorganisation of 1699. Fontenelle's colleagues were hardly quicker than he to appreciate the *Principia*.

mathematician but in Voltaire, whose *Philosophical Letters* (1734) followed by two years the support given to Newton by Maupertuis in a *Discourse on the various Shapes of the Stars*. Attraction, the latter recalled afterwards,

> was emprisoned on its island, or if it crossed the sea, appeared as a monstrosity which had just been condemned. Men congratulated themselves on having banished occult qualities from philosophy, and they had such fear of their return that everything in the least reminiscent of them was a source of fear. They so rejoiced at introducing into philosophy the semblance of mechanism, that they rejected unheard the mechanism that was set before them.[8]

The case for Newton—which became steadily a case against Descartes—could not be wholly ignored, as it was set out with increasing vigour in Richard Bentley's *Confutation of Atheism* (1693), Samuel Clarke's Latin translation (1697) of Rohault's *Traité de Physique* of a quarter century before in which the translator's Newtonian footnotes refuted the Cartesian text, John Keill's *Introduction to true Physics* (1700), David Gregory's *Elements of physical and geometrical Astronomy* (1702), and Roger Cotes' Preface to the second edition of the *Principia*, not to mention the Latin edition of *Opticks* (1706). All but the first of these was in Latin, and so accessible to foreign readers ; Cotes in particular launched out strongly—so strongly that Newton refused to read his Preface before publication, so as to avoid entanglement—against Cartesian insinuations that the Newtonian system was esoteric. The physics and cosmology of Descartes, which in recent memory had seemed so fresh, invigorating and promising in their assertion of pure mechanism, were now in turn pushed on the defensive.

To defend Cartesianism mathematically or experimentally was almost impossible, and few attempted the vain task. (Newton's continental critics did crow over the mathematical errors in the first edition of the *Principia*, corrected in the second ; these did not affect the central issue of celestial mechanics, however.) Cartesian strategy, therefore, dictated a counter-attack on broader principles, through the claim that Descartes represented the genuine tradition of mechanical philosophy, vitiated by Newton's concept of attraction and reintroduction of the cosmic vacuum. How could gravity be anything but a mechanical force, a result of billiard-ball collisions ? How could light be transmitted across empty space ? To the Cartesians the philosophy of the *Principia* as they understood it was not a scientific advance ; it was a retreat to the pre-Cartesian mysteriousness of nature. In their own eyes they were the true progressives, not reactionaries resisting progress.

Nevertheless, the endeavour to defend the vortices *scientifically* yielded only absurdities. Among them, and arousing the contempt of such firm anti-Newtonians as Leibniz and Johann Bernoulli, the abbé Villemot revived in Descartes' name the Ptolemaic spheres (1707) with the hypothesis that each planet was carried round by a separate layer of the aetherial vortex whose velocity was proportional to the square root of its distance from the sun. He absurdly thought that this relation was in accord with Kepler's Third Law.[9] The absurdity was emulated by the philosopher Malebranche (1712) who, realising that Villemot's hypothesis implied that the centrifugal force was constant at all distances from the sun, mistakenly thought that this was proved by the relation $v^2r = k$, which he derived from Kepler's law.[10] He did not notice that this expression was quite different from Villemot's, and supposed that the aether, swirling in a confined vortex, created an inwards pressure equal at all distances from the centre. Compared with this comedy of error the criticism of the astronomer Jacques Cassini (1677–1756), that the Earth is shaped like a lemon rather than a turnip (as Newton and Huygens had correctly reasoned), since it was merely based on a faulty geodetic survey, was at least rational.

It was safer to criticise the *Principia* (as the reviewer in the *Journal des Sçavans* had done in 1688) on the grounds that however accurate and important the book was as a contribution to celestial mathematics, as a contribution to celestial physics it was unintelligible because the mathematical theorems could be given no physical meaning. Newton himself had said that he did not know what gravity was. But the efforts of the French Cartesians at this level too testify rather to their prejudice and lack of vision, than to the logic proclaimed as the strength of their philosophy. The overrated efforts of Malebranche, who had spent his scientific energies since 1674 in adjusting the successive editions of his *Recherche de la Vérité* to the discoveries of true scientists, are typical of the rest. When, in 1712, he mustered arguments against Newtonian physics, he could only harp on the unintelligibility of attraction. It was evident to sense, he thought, that bodies were moved by impulse, and no experiment ever demonstrated that they were moved by attraction, for :

> one plainly sees, in those experiments that seem fittest for proving this kind of movement, that what seemed to be done by attraction is only done by impulse, once one discovers the true and certain cause.[11]

That is, when one decides to beg the question by assuming for true what is to be proved. . . .

Now Newton had never embraced the hypothesis of attraction. In the first edition of the *Principia* he had made it plain that when he used the word, it was to be taken as a " manner of speaking ". In a letter to Bentley (January 1693) he wrote :

> You sometimes speak of Gravity as essential and inherent to Matter. Pray do not ascribe that notion to me ; for the Cause of Gravity is what I do not pretend to know, and therefore would take more Time to consider of it.

And in a further letter he added :

> That Gravity should be innate, inherent and essential to Matter, so that one Body may act upon another at a Distance thro' a Vacuum, without the Mediation of anything else, by and through which their Action and Force may be conveyed from one to another, is to me so great an Absurdity, that I believe no

Man who has in philosophical Matters a competent Faculty of thinking, can ever fall into it.[12]

Did Newton mean that, after all, gravity was a direct mechanical effect produced by the action of the aether, in some such way as Descartes, Huygens, Fatio de Duillier and others had proposed ? It is impossible to think so. The Cartesian aetherial vortices had been thoroughly exposed as fallacies in the *Principia*, and with them the whole structure of Cartesian physics collapsed, incapable of rational revival. It is true that Newton toyed with aetherial hypotheses all his life and that he would, when tormented by charges that he was a mere mathematician, that he believed in esoteric attractions, or that his ideas were physically unintelligible, feign an aetherial hypothesis to show that his discoveries could be accounted for in that way by those who found comfort in such figments for their incapacity to pursue the mathematical way. In the *Letter to Boyle* of 1679 Newton had applied such an aetherial hypothesis to gravity, though this was not yet known to the world at large. He hinted at it again in Query 21 of *Opticks*. But he always made it plain that such hypotheses, like the aetherial hypotheses he fabricated to account for the properties of light, were composed because the " heads of the virtuosos " seemed to require them. Newton was well aware that his own aetherial hypotheses demanded a force of repulsion between aether-particles no less mysterious than the force of gravitational attraction, that an aether freely pervading matter could not affect matter, and that a super-aether would be required to account for the motions of the aether-particles impelling material particles, and so *ad infinitum*.

The Newtonian philosophy of nature postulated that it was enough to define mathematically the action of the forces of nature, such as gravity, in order to explain fully all the effects they produce in phenomena. The method, the example and the proof were displayed in the *Principia*. Was it possible to go further and say what the " causes " of the natural forces are ? Newton himself did not succeed in this, and if his aetherial hypotheses be regarded as

" tubs to whales " it is not clear that he believed the attempt to find such causes could be fruitful. In his third letter to Bentley, just before the passage last quoted, he declared

it is inconceivable, that brute Matter should, *without the Mediation of something else, which is not material*, operate upon, and affect other Matter without mutual contact. (Italics added.)

This and other passages have suggested that for Newton the ultimate nature of forces such as gravity was spiritual. This is true in the sense that he did not believe in billiard-ball mechanisms as the final analyses of physics, but not in the sense that he believed the natural forces of gravity, electricity, chemistry and so on to be miraculous. The law of gravity was for Newton no more a miracle than was the impenetrability of matter. When he wrote that gravity was not innate in matter " in the sense of Epicuros " he meant that neither matter nor gravity were independent of God, who need not have created matter in the first place, or could have created matter without attaching gravity and other physical forces of nature to it. He could have created an inverse-cube law had he pleased. The existence of matter and the action of its forces alike depended on the continuous exercise of the divine will so that the cause of gravity, hidden in the divine plan of things, was as inscrutable as the cause of matter. For the cause of all the fundamental properties and laws of nature was to Newton the divine purpose that willed them ; or (as Newton came near to saying in the General Scholium) God is Nature, though not nature alone.

For Newton this was not a reversible proposition ; it was the eighteenth century that held that Nature is God. Nevertheless his metaphysics—which are perhaps impossible to defend—were strongly attacked by Leibniz and equally warmly championed by Samuel Clarke (1676–1729), critic of Rohault and translator of *Opticks*. Their epistolary debate of 1715–16, in which Clarke was furnished with arguments by Newton behind the scenes, was published in 1717. It was prompted by the second edition of the *Principia* (1713) with Cotes' aggressive Preface and Newton's addition of the General Scholium. The former had already dealt with Leibniz's criticism (1710) that for the planets to move in orbits without the impulse of a medium would be a perpetual miracle. Leibniz now repeated it. He also accused Newton of supposing that the universe does not depend on God ; of weakening natural religion ; of reviving the absurdity of the cosmic vacuum ; and of denying that God had the foresight to make the universe perfect :

> Nay, the machine of God's making is so imperfect, according to these gentlemen, that he is obliged to clean it now and then by an extraordinary concourse, and even to mend it, as a clockmaker mends his work.[13]

The second and third points Clarke simply denied. The fourth he defended by adducing the experimental evidence in favour of the existence of the vacuum. On the fifth he counter-attacked, with the charge that Leibniz's views would exclude God and divine providence from the universe altogether. Leibniz, in short, was a philosopher of that pure mechanism that led to atheism. To the first and repeated assertion that Newton's theory of gravity invoked a constant miracle, Clarke responded by making a distinction between *natural* operations and *mechanical* ones :

> the means by which two bodies attract each other may be invisible and intangible, and of a different nature from mechanism ; and yet, acting regularly and constantly, may well be called natural. . . . If the word *natural forces* means here mechanical, then all animals, and even men, are as mere machines as a clock. But if the word does not mean, mechanical forces ; then gravitation may be effected by regular and natural powers, though they be not mechanical.[14]

Rather unsubtly—for the tendency towards animism is unfortunate —Clarke's opinion seems to echo Newton's : the cause of gravity is something neither purely mechanical, nor purely miraculous.

These discussions of gravity, and of other aspects of Newton's scientific thought, penetrate to the roots of the philosophy of seventeenth-century science. They serve also as a prelude to the philosophy of the next century. Leibniz, himself a mechanist and a great mathematician, opposed nearly as much to Descartes as he was to Newton, demanded of Clarke a resolution of the intellectual dilemma of the seventeenth century : if the universe is wholly mechanical God has no part in it ; if God controls the universe, then it is not mechanical nor governed by laws. It was a dilemma that Leibniz, like Descartes before him, thought he had resolved ; but Newton's mathematical physics rejected his solution no less than Descartes'. Leibniz combined the charge that Newton had lessened the power of God and undermined religion with the apparently contrary criticism that he had also subverted the mechanical philosophy. He invited Newton's spokesman, Clarke, to choose : either the universe is mechanical, or it is miraculous. Clarke (and Newton), like all their forerunners in scientific philosophy, refused such an odious dilemma. But, as Alexandre Koyré has so lucidly explained, history was to prove that Leibniz was right.[15] The role Newton ascribed to God was incompatible with the remainder of his system.

Leibniz, though he perceived the full force of mathematical physics no better than the Cartesians, could not accept a mathematical equation as the ultimate in human knowledge. Newton himself was almost afraid of this, hardly daring to proclaim that beyond the mathematical principles of natural philosophy there is no further understanding ; yet such, after all, is the implication of his statements on gravity. And Newton was clear that in the shadowy realm beyond physics, if there were such, whatever ideas were formed were shaped not by scientific but by religious perceptions.

There can be no doubt that Leibniz saw in Newton's science a serious challenge to sound philosophy and, unlike Malebranche for example, responded to it seriously. This, as he saw, was a moment when the human mind adopted a new course. Paradoxically his debates with Clarke did not sway opinion against Newton ; rather they enhanced the importance of the *Principia* because Leibniz did not treat the book as a mere mathematical treatise. The image of Newton as a philosopher began to appear on the continent, as Locke, Bentley and Clarke had created it in England. The trend was reinforced by a growth of interest in the English empiricist philosophy, and by French admiration for English government and society after the failure of the polity of Louis XIV. The authority of Descartes began to yield gradually on the continent after about 1720. With the acceptance of the *Principia* by the continental mathematicians soon after 1730, the way was open for the eighteenth-century resurgence of mathematical physics on Newtonian principles, in the hands of Clairaut, D'Alembert, Euler, Lagrange and Laplace. But the mathematical was not the only tradition of science that the eighteenth century inherited from Newton, though it was the more important ; nor did the eighteenth century read only the *Principia*. For the author of *Opticks* was also held in high esteem, as the pre-eminent exponent of experimental science and, less happily, as the father of the hypotheses outlined in the *Quaeries*.

As early as 1720 Willem Jakob 'sGravesande (1688–1742) published at Leiden a work entitled *The Mathematical Elements of Physics confirmed by Experiments, or an Introduction to the Newtonian Philosophy*. Recently appointed to the chair of mathematics and astronomy, 'sGravesande had travelled to England a few years before, where he had been converted to Newtonian science, had met Newton, and had been elected to the Royal Society. In spite of the wide influence of Cartesianism in Holland, experimental and observational science flourished there, exemplified by the work of Huygens, Leeuwenhoek, and the successful Dutch anatomical schools. Herman Boerhaave (1668–1738), a powerful figure in Leiden since the first years of the century and a fervent admirer of Bacon's empiricism, was a vocal advocate of experiment in the medical sciences ; he too was influenced by Newton. Another of the group was Pieter van Musschenbroek (1692–1761). 'sGravesande's book was not the first published in Holland to praise the Newtonian method of science, but it was the first to embrace Newton's physics with enthusiasm. For the vortices of Descartes 'sGravesande substituted at Leiden what he called " the true causes of the motions of the heavenly bodies, discovered with great sagacity by the famous Newton ".[16] Besides this, he formed a large collection of experimental apparatus, much of it of his own devising, with which he illustrated the principle theorems of physics. He was not in the least opposed to mathematical reasoning and he recognised that Newton was above all a mathematical philosopher ; but it was the experimental basis of recent English science—no less Boyle than Newton—that he stressed. And experimental demonstration had many advantages from a teacher's point of view. Here s'Gravesande was borrowing again from his English journey ; for Francis Hawksbee (who assisted Newton in some experiments) had been teaching science in London with the aid of demonstrations and models for over a decade, and J. T. Desaguliers (1683–1744), a French protestant refugee, had been doing the same at Oxford since 1710.

The Dutch physicists, whose writings were popular in England, too, when they were translated out of Latin, played a large part both in promoting Newtonian (and more generally English) empirical science on the continent, and in transforming mechanics from the most forbidding of mathematical sciences to a pleasant series of demonstrations. They helped also, however, to spread the confused view that the *Principia* and the theory of gravitation were no less founded upon empiricism than was *Opticks*. Though mathematicians were not misled, the non-mathematically-minded, even among good scientists, to whom only *Opticks* was readily accessible, began to appeal to Newton as the greatest of all practical exponents of the experimental method. In this role Newton largely supplanted Boyle, who had been very frequently praised for the same reason in the period 1670–1710.

Thus Newton, not without some strain, was fitted into the Baconian Royal Society tradition of *Nullius in verba*, and made to

Another detail from Greenwich. Flamsteed is shown with one of his pupils (Thomas Weston, Mathematics Master at Greenwich Hospital) and a large mural quadrant. Note the globe and the clock

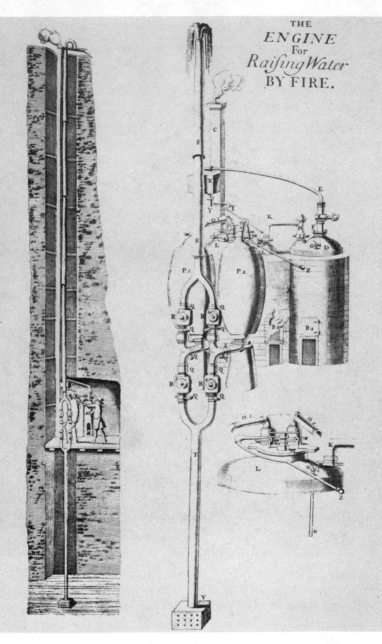

Science applied to technology: Thomas Savery's steam-pump as depicted in the *Lexicon Technicum* of 1704. By manipulating valves, atmospheric pressure forced water up into a vessel where steam was condensed; then, readmitting steam, the water was expelled up the delivery pipe give it further countenance. Provided that other passages (and the very title of the *Principia*) were left out of account, it was not difficult to extract even from that work statements fitting the empiricist interpretation, such as the fourth Rule of Reasoning :

In experimental philosophy we are to look upon propositions inferred by general induction from phenomena as accurately or very nearly true, notwithstanding any contrary hypotheses that may be imagined, till such time as other phenomena occur, by which they may either be made more accurate, or liable to exceptions.

The assessment was more plausible as it came from men interested in problems that could not be handled in the mathematical way. Unfortunately, though Britain produced a number of scientists in the eighteenth century competent to carry on the empirical tradition worthily, there was no one to extend the Newtonian method of mathematical physics. Roger Cotes (1682–1716), from whom (as Newton said) " we might have learned something ", died young. The Scotsman, Colin Maclaurin (1698–1746), was the best of a mediocre bunch. And so it happened that the true development of the Newtonian tradition in theoretical science took place in France and Germany, and was expressed in the language of the Leibnizian calculus, which the empiricist Newtonians of England were neither able to understand nor willing to learn.

For—once again—the British picture of Newton as a theoretical scientist leaned heavily on *Opticks*, while the *Principia* was rather admired than emulated. Everyone quoted Newton's aphorism " I do not feign hypotheses " ; nevertheless empiricists saw that when matters stood in doubt Newton *had* feigned hypotheses, without attending to the cautions with which he had safeguarded himself. They explained the apparent contradiction to themselves by supposing that Newton's speculations were his firm opinions ; as William Whiston put it " when he did propose conjectures in natural philosophy, he almost always knew them to be true at the same time ".[17] Desaguliers, Stephen Hales and others were clearly of the same mind. On this gratuitous assumption Newton's guesses in the *Quaeries* were rebuilt into a new system of explanation, to the point where the pseudo-Newtonian science mimicked the hypotheses of Descartes that Newton had set himself to destroy, just as Descartes, in his time, had tried to root out the errors of Aristotle. In particular the qualitative version of the mechanical (or as one might now say, atomist) philosophy found in the *Quaeries* of *Opticks* was the fount of much further speculation, while the mathematical version and the mathematical treatment of forces in the *Principia* were wholly ignored. Ultimately the empiricist Newtonians in the generation after Newton's death came round to much the same position as that of the Cartesians in the generation before : the *Principia* was a stupendous, immortal, but incomprehensible work of mathematical abstraction ; the reality of physical theory was represented by the qualitative, mechanical, speculative science of the *Quaeries*. Accordingly, they understood Newton's true achievement little better than Malebranche or Leibniz had done.

Chemical ideas, no less than physical hypotheses, were drawn from the *Quaeries*. Newton's own interest in chemistry was at least as old as 1669, when he made a purchase of nitric acid, sublimate of mercury, antimony, alcohol, saltpetre and other materials, and a furnace.[18]

> His Brick Furnaces (as to the manner born) he made and alter'd himself w^{th}out troubling a Bricklayer . . . He very rarely went to bed, till 2 or 3 of ye clock, sometimes not till 5 or 6, lying about 4 or 5 hours, especially at Spring and Fall of ye Leaf, at wch times he us'd to imploy about 6 weeks in his Elaboratory, the Fire scarcely going out either night or day, he siting up one Night as I did another, till he had finished his Chymical Experiments, in ye Performance of wch he was ye most accurate, strict, exact. . . .[19]

So wrote Newton's servitor, forty years afterwards when he could recollect no more than tantalising trivialities of his great master, whose business he never understood. From his fairly well-kept chemical notebook it appears that Newton's preoccupation with chemistry was intermittent through the years at least from 1675 to 1696 ; in the same period he collected, and copiously annotated, both the writings of the scientific chemists like Boyle and those of the " Hermetic philosophers ", the mystery-mongering alchemists.[20] Some useful chemical information was still to be extracted from this dubious source, it is true, but Newton laboured almost equally on materials that now seem utterly worthless from any experimental point of view. His purpose is obscure. He certainly believed, and rightly, that chemistry touches on some of the fundamental properties of matter. Through chemistry knowledge might be won of the forces that hold together, or rearrange, the particles of substances. That Newton was not an alchemist of the type ridiculed by Ben Jonson and long before by Chaucer is quite certain. His chemical notebook contains no trace of the traditional alchemical techniques, though it uses a few of the esoteric alchemical names. Newton called a tin-bismuth alloy "Diana", and a mixture of iron, copper and antimony " the oak ". He wrote (in Latin) :

Neptune with his trident leads the philosopher into the garden of knowledge. Therefore Neptune is a watery mineral solvent and the trident is a ferment of water. . . .[21]

Newton's recorded experiments were nearly all on alloys of metals—often containing antimony, beloved of alchemists—and on attempts to form volatile salts of metals. Perhaps Newton was seeking knowledge of the force that sticks the particles of metals together. Was he trying to make scientific sense of alchemical nonsense ? Not merely to break the alchemists' secret code, as it were, but to discover—if their reports could be verified at all—what rational explanations underlay them ? Newton's temperament inclined him to no love for mysteries and secrets for their own sake, at any rate, and in the end he seems to have decided that the alchemists were a pack of dupes and liars.[22]

Since the mechanical philosophy did not forbid transformations like that of water into earth (accepted by both Boyle and Newton) the transmutation of metals was no impossibility. But it was doubtful indeed whether all imaginable transformations of substance could be effected by chemical means. As Newton put it, if the attraction between particles were inversely proportional to the size of the particle, the smaller particles would be harder to sunder ; so,

the smallest Particles of Matter may cohere by the Strongest Attractions, and compose bigger Particles of weaker Virtue ; and many of these may cohere and compose Bigger Particles whose Virtue is still weaker, and so on for divers Successions, until the Progression end in the biggest Particles on which the Operations in Chymistry, and the Colours of Natural Bodies depend, and which by cohering compose Bodies of a sensible Magnitude.[23]

Of course it was quite obvious that the simple law of gravitational attraction would not work in chemical dynamics. A more complex theory was needed to explain how the principle of attractive force could be applied to solution, distillation, sublimation, double decomposition and all the other motions of particles in chemical reactions. Such a theory, never proposed by Newton himself, was set out in the form of quasi-mathematical theorems by John Keill (1671–1721) as early as 1708.[24] Keill, a mathematician, was a vigorous partisan of Newton ; his paper in the *Philosophical Transactions* was presumably inspired by the latter's long *Quaery* on chemical reaction published two years before in the Latin edition of *Opticks* (1706), which was numbered 31 in the later English editions. Keill, like Newton, supposed that the gross corpuscles of matter were composed of successively more minute orders of particles; that the chemical attractive force varied with the density of a particle of any order ; and that this force decreased as the cube of the distance (or some higher power than the square). From such principles he reasoned about chemical phenomena : for instance a solvent could only dissolve a substance if

1. the particles of the solute are more strongly attracted by those of the solvent than by each other ;

2. the solute has pores permeable by the solvent ;
3. the coherence of the particles of the solute can be destroyed by the violent intrusion of the particles of the solvent.

Keill also endeavoured to link his theorems with those of Newton in the *Principia*. In the following year the physician John Freind (1672–1728) published his *Chymical Lectures : In which almost all the Operations of Chymistry are reduced to their true Principles, and the Laws of Nature*,[25] which also echoed Newton's own language. The " laws of nature " to which Freind referred were Newton's laws, as modified (hypothetically) by Keill. He. considered calcination, distillation, sublimation, fermentation, digestion, extraction, precipitation and crystallization to be the chief processes of chemistry, each of them explicable by the laws of attraction, and he gave many examples. Why do salts dissolve in water, for instance, and not in spirit of wine (alcohol) ? The answer is that

Aqueous particles are more strongly attracted by the Saline Corpuscles, than they are by one another : Whereas in Spirit of Wine, which is indeed much lighter than Water, but more impregnated with Saline Particles, they continue untouch'd. . . .

So again, gold-particles attract mercury very strongly, and the metals amalgamate easily :

But the Attractive Force of Iron and Brass hardly exceeds that of Quicksilver ; for which Reason 'tis, with abundance of difficulty, that they are amalgamated with, unless the Mercury is made to lose some of its Attractive Force, by the Mixture of some other Body.[26]

It was to Newton's own discussion of the applicability of the idea of attractive force to chemistry, however, that early eighteenth-century British chemists most often resorted.[27] They found it in the same famous 31st *Quaery*,* and in another piece of Newton's, *On the Nature of Acids*, published in 1704. In the former Newton asked whether deliquescent salts did not attract water from the air ?

* First published in English in 1717.

Whether the corpuscles of acid were not attracted violently to a metal ? Whether a corpuscle of metallic salt did not consist of a corpuscle of the metal surrounded by a sphere of acid, joined to it by attraction ? Whether, since oil of vitriol was hardly to be separated from water by distillation, yet united at once with a suitable metal, leaving the water,

> doth not this show that the acid Spirit is attracted by the Water, and more attracted by the fixed Body than by the Water, and therefore lets go the Water to close with the fix'd Body ?[28]

And much more of the same kind, in which Newton suggested the play of repulsive as well as of attractive forces : for example, salts though more dense than water do not sink to the bottom, but dissolve evenly :

> does not this imply that the Parts of the Salt or Vitriol recede from one another, and endeavour to expand themselves, and get as far asunder as the quantity of Water in which they float, will allow ? And does not this Endeavour imply that they have a repulsive Force by which they fly from one another. . . ?[29]

Eighteenth-century British chemists, in search of a philosophy and a rationale, pardonably exaggerated the significance of Newton's chemical interest, which was of little long-term effect. Peter Shaw (1694–1763), the translator of Boerhaave's *Elements of Chemistry*, pointing out in 1741 that the content of the *Quaeries* was almost wholly chemical (which is untrue), wrote that it was "by means of chemistry " that Newton had made " a great part of his surprising discoveries "—a silly remark.[30] Like John Freind and Stephen Hales (1677–1761) Shaw hoped to see chemical theory develop on the lines of physical theory. But the eyes of these men was not on mathematics, or the mathematical principles of natural philosophy. They wholly ignored Keill's premature attempt to analyse chemical forces mathematically. Instead they drew their inspirations from the speculations of the *Quaeries*. While flattering themselves that they were following the true experimental method of Newton, and exploiting the idea of attractive force, they forgot that it was by mathematical methods that the attractive force of gravity had been established in the *Principia*.

As with others who have propounded great and difficult ideas, Newton in his last years was ill served by many of those who seemed his best friends. They were almost as dangerous to him as his open enemies. By their violent partisanship they encouraged his impatience of even just criticism and the selfish obtuseness in his nature that blinded him to the real—if not equal—merits of men like Hooke, Leibniz and Flamsteed. Their sycophancy and prejudice impelled him to actions that cannot easily be condoned. Newton's sense of modesty was never strong; his friends behaved as though it were impossible he could have erred, or written a foolish thing. They solemnly accepted his guesses and conjectures—which were shrewd indeed, but not free from mistake and absurdity—as though they possessed the truth of mathematical theorems. Yet these were men from whom Newton carefully concealed his inmost thoughts. The Newtonians obscured a great deal of the subtle complexity that Newton had built into his ideas when, for instance, they insisted that he believed light to be a corporeal emission, or taught that his philosophy of science was purely empiricist. It needed the efforts of Thomas Young (1773–1829), the father of the undulatory theory of light (which he drew from Newton), early in the nineteenth century, to rescue Newton's reputation from the gravestone of hagiographical tradition. Nearly all the faults of the followers of Descartes were repeated by the ardent champions of Newton who, like the Cartesians, preferred materialist hypotheses to the mathematical way that Newton had pursued even more consistently than Descartes. Though eighteenth-century British Newtonians set the *Principia* on a pedestal as the greatest work of the human mind, few had any real insight into the mind that had composed it. When Newton died in 1727, amidst an almost national sense of loss, nothing had happened to advance theoretical physics in Britain a step beyond the level to which he had brought it forty years before. Newton did not live to see his ideas and methods flourishing among the scientists of Europe. He did not therefore have to endure what would surely have been a bittersweet experience for the greatest of theoretical scientists, who was also a proud Englishman ; the brilliant success of mathematical physics abroad, and the richness of a matter-of-fact empirical tradition at home, both in their very different ways looking back to himself as their founder.

EPILOGUE

When Newton died a new age was well begun. Rousseau was a boy of fifteen, Alexander Pope was comfortably settled in the country at Twickenham, Sam Johnson lounged at Oxford, the great Bach was at Leipzig. France and England were, for once, immersed in peace; Russia paused after the headlong reforms of Peter the Great; Holland, Spain, Turkey, the storm-centres of Newton's lifetime, subsided into impotent tranquillity. Wheezing steam-engines pumped the water from Cornish mines. One man was smelting iron in a coal-fired furnace, while another was trying to make rollers do the work of human fingers in spinning thread.

The new age was less tense and more sceptical, less penetrating and more materialist than that into which Newton was born. Privately and publicly men's warmth for religious causes chilled; witchcraft was laughed at as a superstition—except in New England. Torture, devastation by war, perhaps even simple hunger were less cruel than seventy or eighty years before. In Europe, as earlier in England, the iron barriers of class began to soften ever so slightly. Because people were just a little further from starvation their numbers began to multiply without catastrophe, though the death-rate was still murderous. Trade flourished and interest-rates declined. With surprising regularity great ships made the two-year return voyage to India, while smaller ones shuttled back and forth between Europe and the West Indies, New Orleans, Boston and Quebec. For those born at the right levels it was a time of comfort, elegance and complacency.

What did science contribute to the changes that took place in the lifetime of Isaac Newton? Directly, very little. Measuring the speed of light, computing the true shape of the Earth, observing *Volvox* and *Stentor*, added nothing to the welfare of mankind. Apart from a few instrument-makers, science had not given employment to those who needed bread, nor had it provided more bread to be eaten. It had not made diseases less deadly nor cities less pestilential (it was of the coal-smoke of London, offending his nose and eyes, that the great virtuoso John Evelyn had complained, not of its more fatal sewers and water-supply). Inoculation against smallpox, at best a doubtful practice, was borrowed from the peasants of Turkey, and if plague ceased to be endemic the credit lies not with the physicians but with the brown rat. On the other hand, science was no more effective in the arts of death than those of life, since efforts to make the shooting of cannon a mathematical art had proved abortive.

For the best part of two centuries kings, merchants and navigators had looked hopefully to science to tell them how far East or West a ship had sailed—how far North or South was easy to discover. No one doubted that this was a practical problem, no one could deny that it was a scientific one. Even Newton had been called in to judge some proposals for solving it, though he declined to attempt the solution himself. In fact there were no secrets in the matter at all—astronomers knew perfectly well how to measure longitude at sea. What could not be overcome were two practical difficulties. The first was the collection and reduction to tables of a vast number of lunar observations. The second was the training of seamen to use elaborate instruments (of which a prototype was devised by Newton) and to interpret correctly the results thus obtained. In the modern world these difficulties would be got over in a few years ; in the more chaotic world of 1680 or 1730 they were insuperable. There was nothing wrong with science in a case like this ; the deficiency was in the organisation of science. At this time no scientific problem, even one whose solution was easily attainable in principle, could be solved if to do so involved prolonged, expensive and concerted social effort. Hence it was left to the clockmakers, who could work on their own, to bring to success—with minimal help from science—their alternative way of measuring longitude by keeping time accurately.

The social organs of science, the scientific societies, had little power to make science useful. Including as they did men of high talent, and more than talent, their encouragement and recognition is deservedly linked with the finest intellectual achievements of the seventeenth century. What they did affected individuals, for their power over imagination was real. But they were powerless to move society, or to alter the economic and technological patterns entrenched in it. They had neither the money nor the administrative authority to push through major reforms in industry and agriculture, even if they had possessed the knowledge to direct such reforms. For the most part their members had no wish to undertake any such programme, not because they lacked faith in the ability to transform techniques, but because they were reluctant to associate science with government any more closely than it need be. The state by no means lacked the wish, and in a measure the power, to control for its own ends the activities of manufacturers, farmers and merchants but no one as yet thought of making science in any sense the arbiter of economic regulations. The Royal Society, accepting the prevalent mercantilist theories, gave much thought to agricultural practices and issued some exhortation towards improving them—by cultivating the potato, for instance—without ever supposing that its own function was to do more than collect and evaluate information, whose application was the business of those actively concerned with producing things. Few, if any, who believed that science could contribute to human well-being realised how much effort has to be made before it can do so—effort not spent in making a new scientific discovery but in developing it for use and persuading people to use it.

" But nothing does so plainly show the force of rarefy'd Air," wrote John Freind in his chemical book of 1709, " as the Engine by which Mr Savery has taught us to draw Coals out of the Pits ".[1] In

331

fact, Dr Freind did not well understand the new wonder. If he had looked into Harris's *Lexicon Technicum* (1704) he would have learnt that it was by the rarefaction of water and the condensation of steam that flood-water was drawn out of the pits, in a method so " plain and easy," as Harris wrote, " that 'tis an amazing thing . . . that there is no more use made of it ".[2] Indeed, Savery's invention was never much used, and in the very year of Freind's English version (1712) Thomas Newcomen constructed at Dudley Castle, Warwickshire, the first true atmospheric engine. It had a separate boiler, a cylinder, and a piston connected to a rocking beam with the pump-rod at the other end. Yet Freind's mistake was a natural one, since the primitive steam-engine was the lineal descendant of the laboratory pumps that rarefied air. " The father of the steam engine " would be a better title for Robert Boyle than " the father of chemistry ". Since the pressure of the atmosphere held up a column of water 32 feet high, burst bladders and even cracked the experimenters' thick glass receivers, obviously it could be made to do useful work. The problem was to make cheaply and easily the partial vacuum against which work could be done. The cupping-glass technique suggested itself : hot air, cooled, made a partial vacuum. Christiaan Huygens tried using gunpowder to provide the heat—a possible but impracticable method. His assistant, Denis Papin, turned in England to using steam instead. Now the problem of making a working steam-engine was one of engineering development only ; to make the steam in the right place, to lead it into a device for converting the pressure of the atmosphere into work, to cool it so that the air could press down, and then to repeat the cycle.

Huygens had abandoned the heat-engine years before. Papin, whose career was in difficulties, also gave it up. The way to make the atmosphere work was found first by Savery, more elegantly (though still inefficiently) by Newcomen. Two practical engineers had turned the tables on Charles II for laughing at the Royal Society's weighing the air. The moral might seem to be that pure research does pay dividends in the end. The truth is a little more complex. In the first place no scientific revolution was needed to bring the steam-engine into existence ; what Newcomen did could have been done by Hero of Alexandria seventeen hundred years before, who understood all the essential principles.[3] In this respect Boyle's air-pump experiments only emphasised long-familiar facts. And if Huygens, a scientist, saw for the first time the possibility of getting work out of the air (or speaking properly, out of heat), it would have remained an idle dream had not two practical inventors pursued it. And even after their limited success, with all the need to drain mines and fens, to bring water to towns and industries, to pump wort in breweries—not to mention the endless usefulness of motive power—the steam-engine remained for two more generations a mere curiosity.

Yet the story of the steam-engine is a happy one compared with others. There was an idea for an automatic loom in the *Philosophical Transactions* to which no one ever paid attention. There was Prince Rupert's scheme for annealing cast-iron cannon in glass-houses, a failure like the project for machine-guns and multiple-shot pistols. There was Cornelius Drebbel's abortive submarine and, still more sad, Sir William Petty's invention of a twin-hulled ship. Its catamaran construction was to be a grand new principle transforming navigation ; the second prototype, *Invention II*, won a bet in a race with the king's packet-boat from Holyhead to Dublin but the third, *Experiment*, launched by the monarch in person, was lost in a great storm at sea in 1666. The fourth vessel, built in Petty's old age, failed to win any glory. Though he claimed that his invention was merely " an opium to stupefy the sense of my sufferings ", Petty jeopardised the royal favour that he had won by his reputed skill in the " mechanicks of shipping ". Charles II took a dislike to the " bottomless " scheme of the " double-bottomed " ship, and the Royal Society displayed careful lack of interest, voting (in 1663) that " navigation being a state concern was not proper to be managed by the Society ". The shipwrights and the Admiralty opposed the new vessel ; in the end it came to nothing.[4]

A like fate befell Galileo's pendulum-clock (a perfectly workable invention, of which he left only a drawing, however), Roemer's scheme for mathematically shaped gear-teeth, designs for improved carriages, and many other inventions more or less scientifically conceived, among them the functional city-plans of Wren and others. The seventeenth century was prolific in technical inventions and in proposals for ameliorating the human lot from which very little resulted. Many of them sprang from the heads of visionaries, some were lunatic, but a few were realised by subsequent ages.

As a factor affecting the prosperity of European man science was as yet negligible indeed, when compared with secular variations in the incidence of disease, shorter-term oscillations in the bounty of the seasons, the lot of war or peace, the enterprise of merchants forcing their way into colonial markets, or the gradual spread of better farming methods from one village to another. For the vast majority of people life unfolded innocent of anything even remotely connected with science, except perhaps when medical treatment was needed. Even the more fortunate gentry, clergy, merchants and shopkeepers were, with exceptions here and there, untouched by anything more up to date than an astrological forecast. Medicine, the most human and the most universal of the sciences, was also that least modified by new ideas and methods. Whatever happened in anatomy, physiology, embryology, medical chemistry and so on there was little change in the way ordinary physicians, surgeons and midwives went about their business, or even in the regime of the hospitals, these places where men and women went to die rather than to be enabled to live. A case history such as this might have been written at any time from the late Middle Ages to the early nineteenth century :

Colic and Jaundiced illness with complications,
5 February, 1647

Mistress Baldwin of Great Staughton, from a great bitterness felt every morning in her stomach, fell to have each other night a long and cruel pain in her right side, which from thence branched itself to her back . . . At last, the jaundice appeared in a very high manner, and whenever it began to abate, yet at the next colic pain the jaundice came violently on again. For the colical fits, which caused intolerable pain in her stomach, I directed oil of sweet almonds, pepper water . . . which she found much good by. Once I gave her two pretty great bullets which no sooner came to her stomach but in a twinkling they gave her ease, and rested very well that night, but at the next her pains were violent again. In all which fits nothing was so sure a remedy as a tobacco clyster [enema] . . . For the jaundice I purged her with one of rhubarb, which wrought put poorly. I caused a vein to be opened in her arm,* and two or three days after to take leeches ; she felt very present ease, but it lasted not . . . I gave her a vomit (which she had refused before) and found much ease after it for three or four days. Then I directed another purge of rhubarb . . . This wrought 31 times. [!] The next day I caused the vein to be opened the second time. Notwithstanding this, both colic and [jaundice] fits returned. Then I directed her to take three spoonfuls of a vomit . . . All this while I gave her every morning a draught of wormwood, white wine wherein sheep's trittles [droppings] were infused, with a pretty quantity of eggshells powdered . . . But this not prevailing I gave her the purging potion again, which now wrought but ten or eleven times. Then I directed her to take for her ordinary drink small beer posset drink . . . To this, and the sheep's trittle drink as before, with a dram at least of the powder following : eggshells powdered one half, of millipedes drowned in white wine or aqua vitae and dried and Gascoyne's powder each half so much.

To this last revolting concoction Dr Symcotts ascribed the perfecting of the cure, but the patient continued to have relapses until ordered to take " 2 spoonfuls of fresh juice of horsedung in a draught of posset drink [made from beer, in which broom and other herbs were

* The writer, being a physician, did not let blood.

335

infused], with eggshells and millipedes steeped in aqua vitae and dried ". And even this was not quite nasty enough to frighten the disease, for the lady ultimately sought fresh advice in London.5 It would be foolish to pretend that this kind of treatment, or that which relied instead on antimonv, calomel and Dover's powders, was scientific or indeed rational medicine. It was superstitious, magical barbarism.

Yet few escaped it, and the rich were no less superstitious than the poor. Childbirth was deadly peril to both mother and child, and high infant mortality ensured that incessant child-bearing only kept the population stable. The most extravagantly fee'd court physicians could only alleviate with opium the agonies of gout, kidney-stone, cancer, or arthriti. incidental to old age ; violent fevers were almost always fatal, and kings and commoners alike submitted to the disgusting routine of catharsis, vomiting and blood-letting that medicine dictated for the preservation of health by " purging the humours ". A newer prophylactic, " taking the waters ", that received the approval of physicians in the late seventeenth century and brought flourishing spas into existence, was relatively agreeable as it provided opportunity for a pleasant social life. Overdressed, greedy men living in ill-heated and ill-ventilated buildings worried endlessly about their tendency to sweat or take chill, the state of their bowels and bladder, their appetite and digestion—distressed by heart-burn, flatulence, nausea and ulcers—their headaches, cramps, sores and chilblains. (In this world of pain an ordinary toothache was lightly passed over.) They dosed themselves heroically with nostrums whose value was measured by their violence, and with laudanum to recover tranquillity. To undergo surgery was more often than not a sentence of death. Perhaps by a bare margin the odds on surviving amputation or cutting for the stone had lengthened between 1600 and 1700, and the torture of the operation had lessened through adoption of swifter, neater techniques. The clinical classification of diseases had been clarified a little, but Galen could have held his own against any seventeenth-century physician in discussion of their causes. And if chemicals displaced woodlice and horsedung in some practice, their object (as always) was to purge the humours. Venereal disease was ubiquitous, hygiene was appalling. Pepys recorded in his diary when he washed his feet, Newton (it is unkindly recorded) slept in his shirt. Privies in the finest houses were noisome holes that poor wretches cleaned out once a year, and courtiers eased nature in dark corners of the palace of Versailles. Judges were infected by the prisoners they condemned, officers by their starveling troops, for typhus along with typhoid, malaria and enteritis was endemic in all parts of Europe. A large fraction of the population suffered perpetually from vitamin deficiency, manifest in the swollen bellies of children, rickets and mild scurvy. The physician's task was a heavy one, and he had little but self-confidence to help him support it. Life was short, death all too familiar.

It is vain to look for material results, so early as the first decades of the eighteenth century, from the scientific developments that had occurred since the time of Galileo. Their intellectual results, however, were already beginning to be apparent. Since Francis Bacon at least science had identified itself with progress ; if material signs of its civilising mission were scant enough, there were already before the end of the seventeenth century many vocal champions of its contributions to intellectual advancement. Granting the high merit of the Ancients in starting from scratch, nevertheless it was clear (they argued) that the Moderns in the age since the Renaissance had infinitely surpassed them in knowledge of every kind. Naturally not everyone was of this opinion, not Sir William Temple (1628–1699), for example ; "Have the studies, the writings, the productions of Gresham College, or the late academies of Paris, outshined or eclipsed the Lycaeum of Plato, the academy of Aristotle, the Stoa of Zeno, the garden of Epicurus ", he demanded, " Has Harvey outdone Hippocrates, or Wilkins, Archimedes ? "[6] There were dramatists and essayists from Shadwell to Swift to voice derision of the pretensions of modern science, which ended in grubbing in

ditches for insects ; there were also, however, those like Molière and
Pope who took the other side in mocking the meaninglessness of the
old verbal learning, and admiring the solid achievements of the new,
experimental method. Incomprehension of the nature of science
was a stiff barrier (" As for Pitty ", wrote Petty, " let it bee applyed
to ye Ignorance, Incapacity & small obstinancy of ye World "), but
in the first half of the eighteenth century the feeling that progress
had occurred, was occurring and could be hastened gathered force.
Science provided the favourite illustrations of this thesis ; and as
it seemed that science had advanced by rationality, the merit of
choosing reason as a guide in other matters was the more commended.

Philosophy, having gone through a long bout with Descartes,
was reluctant to take steps leading to the complete abandonment of
the old familiar universe. For on the continent the radical youth
of the Cartesian system, when it had seemed destructive of every
scholastic dogma, had been metamorphosed into a conservative old
age. It had been—so the Jesuits were now teaching—all a mistake
to suppose God was an Aristotelian : he was a Cartesian. Through
Descartes' accommodation of his ideas to revelation (which caused
him to suppress *Le Monde* during his lifetime) the trinitarian unity
of science, philosophy and religion that Galileo had endangered had
been for a time restored. Cartesian mechanism did not, after all,
detract from the omnipotence of God ; rightly interpreted, it revealed the infinite majesty of the divine foresight and contrivance.
No Italian was as daring as the Frenchman Malebranche (1638–1715)
in reconciling Cartesianism and faith, but for Catholics everywhere
the effect of the trial of 1632, so disturbing to Descartes himself,
had largely passed away ; it was the mathematical view of nature
that had caused the trouble, not the experimental method or the
mechanical philosophy. Logically, therefore, the later, conservative
Cartesians opposed Newton's mathematisation of physics as staunchly
as the Aristotelians had obstructed Galileo's mathematisation of
motion. What they lost, whether Catholic or Protestant, by their

somewhat hollow reconciliation with authority is obvious enough. This is not to imply that Newtonian scientists and philosophers flouted established religion. Quite the reverse. William Whiston (1667–1752), a Newtonian who failed to hide his Unitarianism, paid a heavy price. Newton, for all his irregularity in attending College Chapel, kept his views on the Council of Nicaea and the divinity of Jesus to himself, though as was his habit, he poured out thousands of words on notes and draft essays which he never intended for any eye but his own. Equally hidden were the reasons for Boyle's tenderness of conscience concerning religious oaths, which prevented his accepting the Presidency of the Royal Society. The problems of all these men were strictly theological and, except in such a rare case as Whiston's, did not inhibit their conforming to the social norms. In any case the great majority of scientists in Britain were untroubled Anglicans. The Church of England, powerfully supported in the Royal Society by its bishops, did not insist on a literal interpretation of Scripture, rather it seemed to find in " natural religion " a strong support for its own theology. It provided Newtonian counterparts to Malebranche and other Cartesian apologists in Richard Bentley and Samuel Clarke. Clearly, had the divine providence not been revealed by faith, it would have been discovered by science. And if Burnet saw no incongruity in fitting a geological theory of the Earth's evolution to the Mosaic narrative, nor more did Newton in his attempt to confirm Biblical history from astronomical records.

Though the paths followed by the Cartesians and the Newtonians, or the continentals and the English, in their respective endeavours to use philosophy as a bridge between the superficially antagonistic principles of scientific and religious truth differed in detail, in essence they were identical. Indeed it is a measure of the weakness of such elastic philosophy that it is hard to see what scientific doctrines could *not* have been comfortably accommodated. At this level the crucial distinctions between the mechanism of Descartes and that of Newton became—despite Leibniz—mere adjustable scientific details: or so it seemed for a decade or two at the end of Newton's life. Philosophers like Leibniz and Berkeley who strove to probe into the fuzzy, complacent notion that the formulation of scientific ideas could never be deeply disturbing to either philosophy or religion—a notion to which Locke lent his weight—were widely misunderstood and could easily be misrepresented as reactionary opponents of experimental and mathematical truths. The more popular and less subtle writers, whose influence was growing enormously even before Voltaire joined their number, simply propagated without criticism the scientists' own statements of their methods and objectives, just as they appealed with satisfaction to the irrelevant piety of the scientists' personal lives. And this was possible, even plausible, because the triumph of the mechanical and mathematical outlook on nature had been accompanied by a violent denunciation of materialism in every other respect ; the seventeenth century, claiming and vindicating its right to be thoroughly mechanist and mathematical in each department of science, even the study of the human frame, had voluntarily abdicated the right to extend the same procedure and the same metaphysic to a province that was not its own. The price paid for scientific materialism, which worked, was the renunciation of philosophical materialism such as all condemned in Thomas Hobbes.

It was in a sense a bargain, attested by Descartes, that nature and the animals should be handed over to mechanistic science while God and the soul and all the ultimates rested with philosophers and divines. The intellectual world was carved into two provinces, as Pope Alexander VI had divided the terrestrial globe. Dualism had been so long developing—since the thirteenth century at least—that acceptance of it was accomplished without tension. It seemed no more than a corollary of the ancient distinction between matter and spirit. Traditionally in European intellectual history philosophy had subserved theology just as bodily life subserved eternal life ; natural philosophy, science, had been one of its many branches. The final recognition of the autonomy of this branch in Newton's lifetime was the culmination of a long and inherently necessary process. Autonomy was restricted by the very purpose of natural philosophy—or so it seemed—and did not imply independence. Science was regarded as independent neither of philosophy nor of religion, which still claimed, and was granted, final authority. But, just as the autonomy of science gave it freedom to exploit successfully a materialist metaphysic within the harmlessness of its own confined domain of natural phenomena, so the very connections of science with philosophy and religion ensured that, once established, this metaphysic must diffuse beyond the postulated limits of scientific inquiry. For these limits were the products of an illusion, one to which the greater and minor scientists from Galileo to Newton had subscribed no less than the philosophers and men of religion who sooner or later embraced and applauded their ideas.

The former were deluded in imagining that there is in the nature of things a barrier to the materialist metaphysic which men would never be eager or able to cross ; a barrier for ever protecting the psychic and spiritual realm, by its own self-evidence, from the methods and questions only appropriate to the material world. Men like Boyle possessed perfect confidence that science could never embarrass religion as they understood it ; they deceived themselves. But philosophers who acted as mediators between the harshness of mathematical and experimental truths and the naive suspicion of the unsophisticated, like divines who found in the developing mechanism of natural science the strongest argument in favour of natural religion, were the more deceived. For their whole position rested upon the barriers, arising from the eternal duality of matter and spirit, which they had been the first to create. When Newton died in 1727 the duality and the barriers still stood almost unshaken ; indeed, his own efforts were directed rather to preserving than to overthrowing them, and it was only the existence of this dualistic metaphysic, at once spiritualist and materialist, that rendered Newtonian physics acceptable in a fundamentally religious epoch. Yet it was also Newtonian physics that in the end brought about the destruction of dualism, and the barriers to science constructed upon it.

Like everyone else, Galileo had accepted the subordination of the realm of natural knowledge to spiritual authority, though he had suffered in person for his attempt to assert the autonomy of scientific thought within that realm. Descartes had made science an offspring of philosophy, finding the first steps to truth rather in meditation than in experiment or mathematics. Newton's physics was no less dependent on a metaphysical view of nature, but his metaphysics were those of pure, abstract mechanism, expressible in mathematical functions alone. Like all his modern predecessors he did not believe that the association of atomism with atheism was a necessary one, nor indeed that the ancient atomists had been atheists ; nevertheless, in the last resort God was as out of place in the Newtonian universe of number as in the Epicurean universe of matter. The divine providence could not be expressed in an equation. As Newtonian mathematical physics was a far more effective and useful science than a qualitative mechanical philosophy like Descartes', so to the same degree it was a more dangerous companion in a dualistic partnership. For the most part Newton's contemporaries did not perceive that the mathematical principles had revolutionised the relations between science, philosophy and religion; nor that, by mathematising matter there was left not more room for the purely spiritual realm, but less. Newton himself was content—or almost content—to tolerate the confines for mathematical physics that were formerly allowed to mechanistic physics. Perhaps there were properties in matter, and forces in nature, that could not be interpreted mathematically. But this was only, in a sophisticated manner, to let God prevail where man was ignorant of a better explanation. Newton's successors saw no need to accept the same restrictions for themselves.

Adhering to the tradition of dualism, Newton had recoiled in horror from the unfortunate deductions Richard Bentley had seemed to draw from the *Principia*. Gravitational attraction was not innate in matter, he declared, nor was it causeless, as giving matter ability to act at a distance would make it. Without knowing what gravity was, or how it could be accounted for by any satisfactory mechanism, he still felt the need to discern some cause, something real, behind the abstraction of the mathematical laws by which gravity is expressed. For otherwise God would become an abstraction too. As earlier mechanists had postulated the divine clockmaker from the clockwork universe, so Newton was impelled to postulate the divine mathematician from the mathematical universe. To his successors, who found no metaphysical difficulty in supposing that gravity is innate in matter, and to whom the mathematical expression of nature was sufficient in itself, that postulate ceased to be necessary. If God was Newton's final hypothesis, he was rejected by later science as Newton himself was thought to have rejected hypotheses.

BIBLIOGRAPHY AND REFERENCES

BIBLIOGRAPHY AND REFERENCES

Original sources have been largely used throughout. Among the most important are the modern complete editions of the works of seventeenth-century scientists, especially those of Descartes, Fermat, Galileo, Huygens, Kepler, Stevin (in progress), and Torricelli. The older *Works* of Boyle (ed. Thomas Birch, 1744, 1772), Malpighi (1686), Newton (ed. Samuel Horsley, 1779–85), Wallis (1693–1699) and many more have not yet been replaced. The publication of the valuable *Correspondence* of Leeuwenhoek, Mersenne, and Newton is still in progress. Useful for the early period is the *Journal* of Isaac Beeckman. Much of the work of Hooke and other early Fellows of the Royal Society is conveniently assembled in R. T. Gunther, *Early Science in Oxford* (1920–45). Some of the numerous separate editions or translations of seventeenth-century scientific writings are listed below.

The lists also include a selection of secondary studies.
General Works : H. Butterfield, *The Origins of Modern Science* (London, 1949) ; A. C. Crombie, *Medieval and Early Modern Science* (New York, 1959) ; A. R. Hall, *The Scientific Revolution : 1500–1800* (London, 1954) ; N. R. Hanson, *Patterns of Discovery* (Cambridge, 1958) ; A. Wolf, *A History of Science, Technology and Philosophy in the 16th and 17th Centuries* (London, 1950).

CHAPTER I: SCIENCE IN TRANSITION

E. A. Burtt, *The Metaphysical Foundations of Modern Physical Science* (London, 1932) ; M. Caspar (trans. C. D. Hellman), *Kepler* (New York, 1959) ; M. Daumas, *Les Instruments Scientifiques aux XVIIᵉ et XVIIIᵉ Siècles* (Paris, 1953) ; J. L. E. Dreyer, *History of Astronomy from Thales to Kepler* (New York, 1953) ; W. E. Houghton, " The English Virtuoso in the Seventeenth Century ", *Journal of the History of Ideas*, III, 1942 ; F. R. Johnson, *Astronomical Thought in Renaissance England* (Baltimore, 1937) ; E. Kremers and G. Urdang, *History of Pharmacy* (Philadelphia, 1940) ; R. Lenoble, *Mersenne ou la Naissance du Mécanisme* (Paris, 1943) ; R. K. Merton, " Science, Technology and Society in Seventeenth-Century England ", *Osiris*, IV, 1938 ; J. M. Stillman, *Story of Early Chemistry*

(New York, 1924) ; E. G. R. Taylor, *The Mathematical Practitioners of Tudor & Stuart England* (Cambridge, 1954).

Note

1. H. W. Robinson and W. Adams (eds.), *The Diary of Robert Hooke, 1672–80* (London, 1935), 100.

CHAPTER II: THE GALILEAN REVOLUTION IN PHYSICS

M. Clagett, *The Science of Mechanics in the Middle Ages* (Madison, London, 1959); L. Cooper, *Aristotle, Galileo and the Tower of Pisa* (Ithaca, 1935) ; H. Crew and A. de Salvio (trans.) *Dialogues concerning Two New Sciences by Galileo Galilei* (New York, 1914 etc.) ; I. E. Drabkin and S. Drake (trans.), Galileo's *On Motion* and *On Mechanics* (Madison, 1960) ; S. Drake (trans.), Galileo's *Dialogues concerning the Two Chief Systems of the World* (Berkeley, 1953) ; *idem, Discoveries and Opinions of Galileo* (New York, 1957) ; *idem* and C. D. O'Malley (trans.), *The Controversy on the Comets of 1618* (Philadelphia, 1960) ; R. Dugas, *La Mécanique au XVIIe Siècle* (Paris, 1954) ; A. Koyré, *Etudes Galiléennes* (Paris, Actualités Scientifiques et Industrielles Nos. 852–854, 1939) ; *idem, From the Closed World to the Infinite Universe* (Baltimore, 1957) ; *idem,* " Galileo and Plato ", *Journal of the History of Ideas,* IV, 1943 ; *idem,* " Galilée et la Révolution Scientifique du XVIIe Siècle", *Conférences du Palais de la Découverte* (Paris), Série D, No. 37, 1957 ; *idem,* " An Experiment in Measurement ", *Proceedings of the American Philosophical Society,* XCVII, 1953 ; E. Mach, *Science of Mechanics* (London, 1942) ; G. de Santillana, *Dialogue on the Great World-Systems* (Chicago, 1953) ; *idem, The Crime of Galileo* (Chicago, 1955).

Notes (pp. 36–77)

1. *Hamlet,* v, 1, lines 215–6.
2. Robert Boyle, *An Essay of the Great Effects of even Languid and Unheeded Motion* (London, 1685).
3. Andrew Motte (rev. F. Cajori), *Sir Isaac Newton's Mathematical Principles of Natural Philosophy* (Berkeley, 1946), 399.
4. I am not seeking to determine the philosophic status of the laws of motion as self-evident truths, definitions, or empirical facts ; but merely to emphasise that they may be regarded as unfalsifiable.
5. The parallel passages are in S. Drake, *Discoveries :* 28, 57, 90–3, 93–4, 118 ; and *idem, Dialogues,* 46–59, 68–80, 321–7, 340.
6. Galileo, *Opere,* Edizione Nazionale, XIV, 340 ; XVI, 162.
7. Drake, *Dialogues,* 32.
8. Galileo, *Opere,* XI, 344–5. The same letter makes it clear that Galileo accepted the epicyclic motions of Copernicus as adequate representations of

the planetary orbits. In this reply Cesi refers to the elliptical theory of Kepler.

9. Drake, *Dialogues* : 29.

10. *Ibid.*, 37 ; cf. also 18.

11. *Ibid.*, 19, 32.

12. *Ibid.*, 33.

13. *Ibid.*, 132.

14. *Ibid.*, 217.

15. *Ibid.*, 145.

16. *Ibid.*, 248.

17. *Ibid.*, 116.

18. *Ibid.*, 221–9. Of course neither the fourteenth century nor Galileo expressed the relations as generally as does the modern notation.

19. Galileo, *Opere*, XIV, 362, 3 July 1632 ; XVI, 201.

20. *Ibid.*, X, 228–30, 11 February 1609.

21. *Ibid.*, XIV, 171, 3 December 1630.

22. *Ibid.*, XVI, 177, 21 December 1634.

23. *Ibid.*, XVIII, 78, 1 August 1639.

24. *Ibid.*, XVIII, 67–8, 28 June 1639.

25. Crew and de Salvio, *op. cit.*, 71, 62. Aristotle never offered so absurd and categorical an example as Galileo put into his mouth.

26. *Ibid.*, 65.

27. *Ibid.*, 4.

28. Drake, *Dialogues*, 203–8.

29. Crew and de Salvio, 165.

30. *Ibid.*, 162–6. Mach (*op. cit.*, 162–3), who did not represent Galileo accurately, assumed what was to be proved, i.e. that the time and deceleration of ascent are the same as the time and acceleration of ascent.

31. *Ibid.*, 162.

32. Galileo, *Opere*, XI, 85, 9 April 1611. The change of Galileo's opinion may well be much earlier than this reference to it.

33. Crew and de Salvio, 160.

34. *Ibid.*, 171–2.

35. Drake, *Dialogues*, 53–4.

36. *Ibid.*, 51.

37. *Ibid.*, 152–4.

38. Drake, *Discoveries*, 276–7.

39. Drake, *Dialogues*, 328, 335.

40. *Ibid.*, 462.

CHAPTER III: NATURE'S LANGUAGE

C. B. Boyer, *The History of the Calculus* (New York, 1959) ; J. M. Child, *The Early Mathematical Manuscripts of Leibniz* (Chicago, 1920) ; J. F. Scott, *History of Mathematics* (London, 1958) ; D. E. Smith, *History of Mathematics* (Boston, 1923–5) ; H. W. Turnbull, *James Gregory Tercentenary Memorial Volume* (London, 1939) ; *idem, Mathematical Discoveries of Newton* (London, 1945).

Notes (pp. 78–102)

1. Crew and de Salvio, *op. cit.* (Ch. II), 7.

2. Drake, *Dialogues*, 14.

3. *Ibid.*, 103.

4. Drake, *Discoveries*, 237–8.

5. Napier, the inventor of logarithms, published a table of logarithms of sines in 1614 (*Mirifici Logarithmorum Canonis Descriptio*) so calculated that $\log_N a = 10^7 \log \dfrac{10^7}{a}$ (thus, taking the whole sine as 1 in place of 10^7, the base $N = e^{-1}$). Henry Briggs published the common logarithms of the first thousand numbers in 1617.

6. Napier's successors perceived that the relation between the numbers in two progressions, one arithmetic, the other geometric, is logarithmic. Hence logarithms entered the theory of series. Moreover, as the quadrature of certain curves derived from the summation of series, logarithms were found necessary in obtaining the integrals of certain functions.

7. Crew and de Salvio, 32–6.

8. Kepler (*Nova Stereometria Doliorum Vinariorum*, 1615) employed a variety of methods for calculating areas ; Luca Valerio (*De centro gravitatis solidorum*, 1604) a development of Archimedes'. For the relations of Galileo and Cavalieri, cf. *Opere*, XIII.

9. That areas of infinite extent—or the corresponding infinite series—may have a finite limit was established in the fourteenth century, and again demonstrated by Torricelli, Roberval and Fermat in the seventeenth.

10. *The Geometry of René Descartes*, facsimile with trans. by D. E. Smith and M. L. Latham (New York, 1954), 3.

11. In "An Account of the Book entitled Commercium Epistolicum" (*Phil. Trans.*, 1715, p. 206) it is stated : "By the help of the new Analysis Mr Newton found out most of the Propositions in his *Principia Philosophiae* but . . . he demonstrated the propositions synthetically that the system of the heavens might be founded on good geometry". A longer version of this statement is in Newton's hand among his papers (Cambridge University Library, MS. ADD. 3968, no. 13).

12. There is no space to review the independent development of Leibniz's thought, on the basis of the work of Barrow and Descartes, by which he had formulated the principles of the calculus. Child argues cogently that this development was complete, without Leibniz having any familiarity with Newton's discoveries in mathematics, by 1677.

CHAPTER IV: THE METHOD OF SCIENCE

A. G. A. Balz, *Cartesian Studies* (New York, 1951) ; A. Gewirtz, " Experience and the non-mathematical in Descartes ", *Jour. Hist. Ideas*, II, 1941 ; A. Koyré, *Entretiens sur Descartes* (New York and Paris, 1944) ; R. Lenoble, *Mersenne ou la Naissance du Mécanisme* (Paris, 1943) ; G. Milhaud, *Descartes Savant* (Paris, 1921) ; P. Mouy, *Le Développement de la Physique Cartésienne* (Paris, 1934) ; J. F. Scott, *The Scientific Works of René Descartes* (London, 1952).

Notes (pp. 103–31)

1. Descartes, *Discourse on Method*, trans J. Veitch, Everyman Library (London, 1912 etc.), 1. In quoting from this translation I have occasionally modified slightly both the phraseology and the punctuation.

2. Bacon, *Novum Organum*, II, xi–xx.

3. Motte-Cajori, *Principia*, 547. In the context the famous *Hypotheses non fingo* does not indeed signify " I *never* feign hypotheses ", but Newton did go on to say " hypotheses have no place in natural philosophy ". And the same opinion is repeated at the end of *Opticks* : " the main Business of natural Philosophy is to argue from Phaenomena without feigning Hypotheses " (5th ed., 1931, 369).

4. *Discourse on Method*, 5.

5. *Ibid.*, 16.

6. *Ibid.*, 17.

7. *Ibid.*, 27.

8. *Ibid.*, 31.

9. *Ibid.*, 32.

10. *Ibid.*' xiv–xv.

11. Descartes, *Œuvres* (ed. Ch. Adam et P. Tannery, Paris, 1897–1913), IX, 37.

12. *Ibid.*, IX, 102.

13. *Discourse on Method*, 50.

14. *Ibid.*, 51.

15. *Correspondence of Isaac Newton* (ed. H. W. Turnbull, Cambridge, 1959–), I, 204–10, 6 July 1672.

16. *Ibid.*, 110.

17. *Ibid.*, 235.

18. T. Birch, *History of the Royal Society* (London, 1756), I, 180–2. The experiment was made by lowering a slender, weighted glass pipe, closed at the top and open at the bottom, to measured depths in a larger tube filled with water. The depth of submersion required for each successive half-inch compression of the air in the little pipe was noted. Hooke did not record the barometric pressures prevailing at the time of the experiments. I have assumed (from his own reckoning) that it was 29.1 inches of mercury, equal to 33 feet of water, in calculating the theoretical lengths from Boyle's Law. I have adjusted the figures for Experiment III (pressure about 28.3 in. ?) accordingly.

CHAPTER V: FLORENCE, LONDON, PARIS

T. Birch, *History of the Royal Society* (London, 1756); J. Bertrand, *L'Académie des Sciences et les Académiciens de 1666 à 1793* (Paris, 1869); H. Brown, *Scientific Organizations in Seventeenth-Century France* (Baltimore, 1934); W. E. Houghton, "The History of Trades", *Jour. Hist. Ideas,* II, 1941; Sir H. Lyons, *The Royal Society: 1660–1904* (Cambridge, 1944); M. Ornstein, *The Role of Scientific Societies in the Seventeenth Century* (Chicago, 1928 etc.); T. Sprat, *History of the Royal Society* (1667; ed. J. I. Cope and H. W. Jones, Missouri, 1958); R. H. Syfret, "The Origins of the Royal Society", *Notes and Records of the Royal Society,* v, 1948, also *idem, ibid.,* VII, VIII, 1950; C. R. Weld, *History of the Royal Society* (London, 1848).

Notes (pp. 132–54)

1. Robert Hooke, memorandum of 1663, quoted Weld, I, 146–7.
2. G. B. Shaw, *In Good King Charles's Golden Days.*
3. Brown, 75–6.
4. *Ibid.,* 84.
5. *Ibid.,* 101; "Words are feminine, facts masculine".
6. *Ibid.,* 127.
7. The sixteen Academicians first appointed were: Carcavy, Frénicle, Roberval (mathematicians); Auzout, Huygens, Picard (astronomers); Buot (engineer); Gayant, Pecquet, Claude Perrault (anatomists); Bourdelin, Duclos (chemists); Delachambre (royal physician); Marchand (botanist); Mariotte (physicist); and Gallois (editor), with Duhamel (secretary). Thirty-one other academicians were appointed before 1699, among them Blondel, Cassini, de la Hire, Homberg, Lemery, Leibniz, Roemer, Tournefort and Varignon.
8. This nexus of ideas was not by any means unique with the English Puritans. A more permanent monument to it exists in the Scottish Universities.
9. Quoted Weld, I, 30–3, 35–7.
10. Birch, *History,* I, 3.

11. A volume of *Mémoires de Mathématiques et de Physique tirés des Régistres de l'Académie des Sciences* was published in 1692. The volumes of *Mémoires* of the Académie from 1666 to 1699 were printed only in the early eighteenth century.

12. Cf. G. N. Clark, *Science and Social Welfare in the Age of Newton* (Oxford, 1949), 16.

13. L. A. Foucher de Careil, *Œuvres de Leibniz* (Paris, 1859–75), VII, 64–74 ; C. I. Gerhardt, *Mathematische Schriften*, in Leibniz's *Gesammelte Werke* hrsg. von G. H. Pertz, IV, 519.

14. Originally published as *Philosophical Transactions of the Royal Society*, under Robert Plot's editorship the title was temporarily altered to *Philosophical Transactions giving some Accompt of the present Undertakings, Studies and Labours of the Ingenious in many considerable Parts of the World*, breaking the connection with the Royal Society. But papers not by Fellows had always been included.

CHAPTER VI : EXPLORING THE LARGE AND THE SMALL

(a) A. E. Bell, *Christiaan Huygens and the Development of Science in the Seventeenth Century* (London, 1947) ; I. B. Cohen, " Roemer and the first determination of the Velocity of Light (1676) ", *Isis*, XXXI, 1940 ; Daumas, *op. cit.* (Ch. I) ; M. 'Espinasse, *Robert Hooke* (London, 1956) ; B. le B. de Fontenelle, *Entretiens sur la Pluralité des Mondes* (ed. R. Shackleton, Oxford, 1955) ; R. Grant, *History of Physical Astronomy* (London, 1852) ; H. C. King, *History of the Telescope* (London, 1955) ; A. Koyré, *From the Closed World to the Infinite Universe* (Baltimore, 1957) ; E. F. MacPike, *Correspondence and Papers of Edmond Halley* (Oxford, 1932) ; *idem*, *Hevelius, Flamsteed and Halley* (London, 1937) ; D. Shapley, " Pre-Huygenian Observations of Saturn's Rings ", *Isis*, XL, 1949 ; D. J. Price in Singer *et al.*, *History of Technology*, III (Oxford, 1957).

(b) R. S. Clay and T. H. Court, *History of the Microscope* (London, 1932) ; F. J. Cole, *Early Theories of Sexual Generation* (Oxford, 1930) ; *idem*, *History of Comparative Anatomy* (London, 1949) ; C. Dobell, *Antony van Leeuwenhoek and his " Little Animals "* (London, 1932) ; N. Grew, *Anatomy of Plants* (London, 1682) ; R. Hooke, *Micrographia* (1665 ; in R. T. Gunther, *Early Science in Oxford*, XIII) ; A. F. W. Hughes, *History of Cytology* (London & New York, 1959) ; A. van Leeuwenhoek, *Alle de Brieven* (Amsterdam, 1939–) ; M. Malpighi, *Opera Omnia* (London, 1687) ; J. Swammerdam, *Book of Nature* (trans., T. Flloyd, London, 1758).

Notes (pp. 155–74)

1. Auzout to Oldenburg, 2 July 1665, Royal Society MS. A. I, 5.

2. He supposed the distance of Saturn to be a geometrical mean between the radius of the sun (15 times that of the Earth) and the distance of the stars.

3. H. Woolf, *Transits of Venus* (Princeton, 1959), Ch. 1. In *Opticks* Newton makes the parallax 12″, and the solar distance 70.10⁶ miles.

4. Newton, *System of the World*, Motte-Cajori, 596. By a slip in text, the sexagesimal "iv" is given for "iii". Roberts, *Phil. Trans.*, xviii, 1694, 101–103.

5. Copernicus, *De Revolutionibus Orbium Caelestium*, Book 1, end of section 6.

6. Newton, *Opticks* (5th ed., 1931), 261.

7. Fontenelle (*ed. cit.*), 40–5.

8. G. D. Cassini, *Ephemerides Bononienses Mediceorum Syderum* (Bologna, 1668).

9. Cohen, *loc. cit.*

10. Such instruments had been constructed by the Englishman William Gascoigne about 1640 but without any subsequent effect.

11. MacPike, *Halley*, 2.

12. Newton, *Principia*, 1687, 508. Hevelius (*Cometographia*, 1668) had suggested that a cometary orbit is parabolical, and Dörfel had shown that the comet of 1681 satisfied this condition.

13. This is the recurring comet mentioned by Newton in *Principia* Book iii, third edition (Motte-Cajori, 515). In the first two editions Newton did not speak of Halley's work on recurrent comets, nor did he ever refer to the famous "Halley's Comet" of 1682.

14. *Book of Nature*, 1, 79.

15. *Alle de Brieven*, 1, 165.

16. Letter of November 1677, *ibid.* ii, 279 ff. Spermatozoa were first seen by a physician named Ham, who directed Leeuwenhoek's attention to them.

CHAPTER VII: PROBLEMS OF LIVING THINGS

In addition to the works noted for the last chapter, under (b) : F. D. Adams, *Birth and Development of the Geological Sciences* (New York, 1954) ; A. Arber, *Herbals* (Cambridge, 1958) ; M. Foster, *Lectures on the History of Physiology* (Cambridge, 1924) ; G. Keynes, " The History of Blood Transfusion ", *Penguin Science News*, iii, 1947 ; A. O. Lovejoy, *The Great Chain of Being* (Cambridge, Mass., 1948) ; D. McKie, " Fire and the Flamma Vitalis ", in E. A. Underwood (ed.), *Science, Medicine and History* (Oxford, 1953) ; A. W. Meyer, *An Analysis of the De Generatione Animalium of William Harvey* (Stanford University, 1936) ; M. F. A. Montagu, *Edward Tyson* (Memoirs of the American Philosophical Society, xx, Philadelphia, 1943) ; J. Needham, *History of Embryology* (Cambridge, 1934) ; E. Nordenskiold, *History of Biology* (New York, 1946) ; J. R. Partington, " Life and Works of John Mayow ", *Isis*, xlvii, 1956 ; T. S. Patterson, " John Mayow in contemporary setting ", *Isis*, xv, 1931 ; C. E. Raven, *John Ray* (Cambridge, 1950) ; *idem, English Naturalists from Neckham to Ray* (Cambridge,

1947) ; *idem, Natural Religion and Christian Theology* (Cambridge, 1953) ;
F. Redi, *Experiments on the Generation of Insects* (1668), trans. M. Bigelow (Chicago,
1909) ; J. Sachs, *History of Botany*, trans. H. E. F. Garnsey and I. B. Balfour
(Oxford, 1890) ; N. Steno, *Prodromus*, trans. J. G. Winter (New York, 1916) ;
C. Schneer, "Rise of Historical Geology in the Seventeenth Century ", *Isis*,
XLV, 1954.

Notes (pp. 175–215)

1. W. Charleton, *Enquiries into Human Nature* (London, 1680), 390.

2. This is the traditional number of known plants at the end of the seventeenth
century. Raven (*Ray*, 241) says that Ray " knew and treated as distinct " more
than 6,000, but listed far more.

3. *Methodus Plantarum*, Preface ; quoted from the translation of Raven,
Ray, 193.

4. *Historia Plantarum*, I, 1188 ; quoted from *ibid.*, 234.

5. John Lowthorp, *Phil. Trans. Abridged to 1700*, II, 425–6.

6. Hooke, *Micrographia* (1665), 111.

7. Swammerdam, *Book of Nature*, II, 69.

8. H. Power, *Experimental Philosophy* (1664).

9. Swammerdam, *op. cit.*, I, 3.

10. *Ibid.*, II, 104.

11. Ch. Bonnet, *Considérations sur les corps organisés* (Amsterdam, 1762),
quoted by Cole, *Early Theories of Sexual Generation*, 97–8.

12. R. de Graaf, *De mulierum organis generationi inservientibus* (Leyden, 1672).

13. *Alle de Brieven*, II, 277 ff., November 1677 ; *Phil. Trans.*, no. 142,
1678.

14. *Ibid.*, II, 335, 18 March 1678 ; IV, 11–19, 22 January 1683 ; IV, 57–66,
16 July 1683.

15. *De l'Homme* is the second part of *Le Monde* ; it was first printed in Latin
in 1662, in French in 1664.

16. Descartes, *Œuvres* (ed. Ch. Adam et P. Tannery), XI, 131–2.

17. *Ibid.*, 202.

18. Nils Stensen (Steno), *De musculis observationum specimen* (1664) ; *Elementorum myologiae specimen* (1667).

19. Vesalius, *De humani corporis Fabrica* (1543), quoted Foster, *op. cit.*, 68 ;
W. Harvey, *De motu locali animalium* (Cambridge, 1958), 89, 117.

20. G. A. Borelli, *De motu animalium* (Rome, 1680–1), II, 56–65.

21. *Ibid.*, II, 226.

22. J. B. Helmont, *Ortus Medicinae* (Amsterdam, 1648) ; English trans. by
J. Chandler, *Oriatrike* (London, 1662).

23. R. P. Multhauf, "J. B. van Helmont's Reformation of the Galenic Doctrine of Digestion ", *Bulletin of the History of Medicine*, XXIX, 1955, 161.

24. H. Guerlac, "John Mayow and the Aerial Nitre ", *Actes du 7ᵐᵉ Congrès Int. d'Hist. des Sciences*, 1953, 332–49.

25. G. Ent, *Apologia pro Circulatione Sanguinis* (1641), quoted *ibid.*, 342.

26. *Micrographia*, 103.

27. T. Birch, *History of the Royal Society* (London, 1756), II, 198 ; 10 October 1667.

28. R. Lower, *Tractatus de corde*, 1669, 168–9. R. T. Gunther, *Early Science in Oxford*, IX, with English trans. by K. J. Franklin.

29. J. Mayow, *Tractatus quinque medico-physici* (1674) included revised versions of his *Tractatus duo* (1668). A treatise on respiration appeared in both.

30. *Idem, Medico-Physical Works*, Alembic Club Reprints, 17. (Edinburgh, 1907), 205, 206.

31. The claim by Foster and others that Mayow had discovered oxygen deserves no further refutation.

32. Hooke made an experiment with an animal enclosed in a box, in which the air was re-circulated by bellows. Unfortunately he concluded that this fanning had a marked effect on prolonging the life of the animal.

33. E. Tyson, *Orang-Outang, sive Homo Sylvestris : or, the Anatomy of a Pygmie* (1699), 91. The word chimpanzee was first used in English in 1738. The skeleton of Tyson's specimen is still preserved in the British Museum, Natural History.

34. J. Addison , *Fragments, or Minutes of Essays ; Works*, (1809), VIII, 231 ; quoted by Lovejoy, *op. cit.*, 196.

35. J. Locke, *Essay concerning Human Understanding*, III, vi, sec. 12 ; quoted *ibid.*, 184.

36. A. Pope, *Essay on Man*, Ep. II, lines 3–10 ; *ibid.*, 199.

CHAPTER VIII: ELEMENTS AND PARTICLES

M. Boas, " The Establishment of the Mechanical Philosophy ", *Osiris*, X, 1952 ; *idem, Robert Boyle and Seventeenth-Century Chemistry* (Cambridge, 1958) ; *idem,* " Structure of Matter and Chemical Theories in the 17ᵗʰ and 18ᵗʰ Centuries " in *Critical Problems in the History of Science* (Madison, 1959) ; *idem* and A. R. Hall, " Newton's Mechanical Principles ", *Jour. Hist. Ideas*, XX, 1959 ; *idem,* " Newton's Chemical Experiments ", *Archives Int. d'Hist. des Sciences*, XI, 1958 ; M. B. and A. R. Hall, " Newton's Theory of Matter ", *Isis*, LI, 1960 ; F. Hoefer, *Histoire de la Chimie* (Paris, 1866) ; J. Mayow, works cited in notes 24 and 30, Ch. VII ; H. Meztger, *Les Doctrines Chimiques en France du début du 17ᵉ à la fin du 18ᵉ Siècle* (Paris, 1923) ; J. R. Partington, *Short History of Chemistry* (Cambridge, 1951) ;

idem, "Jean Baptista van Helmont", *Annals of Science,* I, 1936 ; Jean Rey, *Essays* (1630), ed. D. McKie (London, 1951) ; J. M. Stillman, *Story of Early Chemistry* (New York and London, 1924).

Notes (pp. 216–43)

1. J. Locke, *An Essay concerning Human Understanding,* Book II, ch. xxiii, sec. 11.
2. S. Drake, *Discoveries and Opinions of Galileo,* 277–8.
3. F. Bacon, *Novum Organum,* II, xxiii.
4. R. Boyle, *Works* (1772), I ; 356.
5. J. R. Glauber, *Works* (trans. C. Packe, London 1689), Part I, 123.
6. *Ibid.,* 251.
7. Boyle, *Sceptical Chymist,* Propositions II and III.
8. Boyle, *Experiments and Considerations touching Colours* (1664), 40.
9. *Ibid.,* 316 ; *Origine of Forms and Qualities* ; Boyle, *Works* (1772), III, 83–5.
10. *Ibid.,* 105 ; P. Shaw, *Boyle's Works Abridged,* I, 267.
11. Boyle, *Colours,* 302–8.
12. Hooke, *Micrographia,* 103 (abbreviated).
13. Mayow has been praised for the cogency of his experiments. But in arguing from combustion over water that the elasticity of the residual air was reduced, he ignored Boyle's demonstrations that the pressure in an enclosed volume of dry air is not changed by burning.
14. Newton, *Opticks,* 5th ed. (1931), 400. Some of the quotations in the following pages are taken from A. R. and M. B. Hall, *Unpublished Scientific Papers of Isaac Newton* (Cambridge, 1962).
15. Newton, MS. *Conclusion to Principia.*
16. *Idem,* MS. draft of Preface to *Principia.*
17. *Idem,* MS. *Conclusion.*
18. *Opticks,* 380–1.
19. Newton, MS. draft of Preface.
20. *Principia,* ed. Motte-Cajori, 547 ; but see further M. B. and A. R. Hall, "Newton's Electric Spirit : Four Oddities", *Isis,* L, 1959.

CHAPTER IX : EXPERIMENTAL PHYSICS

E. N. da C. Andrade, "Robert Hooke", *Proceedings of the Royal Society,* A, CCI, 1950 ; A. C. Crombie, "Newton's Conception of Scientific Method", *Bulletin of the Institute of Physics* (1957) ; I. B. Cohen (ed.) *Isaac Newton's Papers and Letters on Natural Philosophy* (Cambridge, Mass., 1958) ; A. R. Hall, "Sir Isaac Newton's Note-book, 1661–65", *Cambridge Historical Journal,* IX, 1948 ; *idem,* "Further Optical Experiments of Isaac Newton", *Annals of Science,* XI, 1955 ; C. Huygens,

Treatise on Light (trans. S. P. Thompson, London, 1912) ; A. Koyré, " L'hypothèse et l'expérience chez Newton ", *Bull. Soc. française de Philosophie*, 1, 1956 ; V, Ronchi, *Histoire de la Lumière* (trans. J. Taton, Paris, 1956) ; I. H. B. and A. G. H. Spiers, *The Physical Treatises of Pascal* (New York, 1937) ; C. de Waard, *L'Expérience Barométrique* (Thouars, 1936).

Notes (pp. 244–75)

1. Newton, *Opticks*, 404.

2. Newton—against Descartes—exposed the conservation of motion as a fallacy. It was inconsistent with, and made superfluous by, his own concept of force.

3. *Principia*, Motte-Cajori, 399.

4. Crew and de Salvio, *op. cit.*, (Ch. II), 94–102.

5. If Galileo concluded that frequency varies inversely with the length of the vibrating string, then these ratios of pitch would follow at once from the known corresponding ratios of string-length, i.e. $1, \frac{3}{4}, \frac{2}{3}, \frac{1}{2}$.

6. Crew and de Salvio, 16.

7. Spiers, 99 ; 15 November 1647.

8. *Ibid.*, 75.

9. *New Experiments Physico-Mechanical touching the Spring of the Air and its Effects* (1660).

10. *Ibid.*, *Works*, I, 11.

11. Boyle discovered his law from experiments on the compression of air, independently of Richard Towneley who discovered it from others on expansion. The latter were also performed by Hooke, who acknowledged the priority of Towneley's " hypothesis ". The claim of Edmé Mariotte — the best of the French physicists—to eponymity with Boyle is ill-founded, for the law was discussed by him only in 1676 (*Oeuvres de Mariotte*, Leiden, 1717, I, 151).

12. Historians have expressed sympathy for Hooke, as though in later years he was pursued by Newton's enmity. This judgement is false. Hooke had indeed the double misfortune, in his first encounter with Newton, to oppose a man stronger than himself, and to be in error. He set out deliberately—for if the terms of his report on Newton's paper were not measured they were idiotic—to destroy it by asserting in the strongest language that the experiments were unoriginal and the conclusions unacceptable. If, after this, Hooke made conciliatory moves (for his onslaught upon Newton did not commend itself to many Fellows of the Royal Society), he nevertheless tried on two further occasions to make Newton appear both a fool and a plagiarist. It was only after this that Newton ceased to be civil to Hooke.

13. *Micrographia*, 64.
14. Newton to Oldenburg, 18 January 1672. *Correspondence*, I, 82–3.
15. *Ibid.*, 96–7.
16. *Ibid.*, 164.
17. *Papers and Letters*, 136 ; *Correspondence*, I, 290–5.
18. *Correspondence*, I, 111.
19. The *relative* wavelengths in Newton's "fits" theory are much more accurate. According to modern measurements the wavelengths of violet, blue, green, yellow, orange and red light are about as 1, 1·13, 1·25, 1·40, 1·45, 1·60 ; according to Newton's harmonic scale the relative lengths of the "fits" are about as 1, 1·11, 1·21, 1·30, 1·35, 1·50.
20. *Opticks*, Quaery 26 (5th edn., 360).
21. *Ibid.*, 244.

CHAPTER X : NEWTON AND THE WORLD OF LAW

Besides many works on Newton already mentioned : W. W. R. Ball, *An Essay on Newton's Principia* (London, 1893) ; D. Brewster, *Memoirs of Sir Isaac Newton* (Edinburgh, 1855) ; W. J. Greenstreet (ed.), *Isaac Newton, 1642–1727* (London, 1927) ; A. R. Hall, "Newton on the Calculation of Central Forces", *Annals of Science*, XIII, 1957 ; History of Science Society, *Isaac Newton* (London, 1928) ; A. Koyré, *op. cit.*, Ch. VI (a) ; *idem*, "Gravitation Universelle de Kepler à Newton", *Conf. du Palais de la Découverte*, 1951 ; *idem*, "Significance of the Newtonian Synthesis", *Arch. Int. d'Hist. des Sciences*, XI, 1950 ; *idem*, "Mécanique Céleste de J. A. Borelli", *Rev. d'Hist. des Sciences*, V, 1952 ; *idem*, "Unpublished Letter of Robert Hooke to Isaac Newton", *Isis*, XLIII, 1952 ; *idem*, "Documentary History of Fall", *Trans. Amer. Phil. Soc.*, N. S., IV, 1955 ; L. T. More, *Isaac Newton* (New York, 1934) ; L. D. Patterson, "Hooke's Gravitation Theory and its Influence on Newton", *Isis*, XL, 1949 ; Royal Society, *Newton Tercentenary Celebrations* (Cambridge, 1947); E. W. Strong, "Newton's Mathematical Way", *Jour. Hist. Ideas*, XII, 1951 ; *idem*, "Newton and God", *ibid.*, XIII, 1952 ; R. S. Westfall, *Science and Religion in 17ᵗʰ Century England* (London, 1958) ; *idem*, "Isaac Newton : Religious Rationalist or Mystic ?", *Rev. of Religion*, XXII, 1958.

Notes (*pp. 276–306*)

1. *Principia*, Motte-Cajori, 12.
2. Cambridge University Library, ADD. MS. 3968, no. 2.
3. Huygens' propositions on centrifugal force were as yet unpublished ; Newton's method was briefer but less formal.

4. Hall, *loc. cit.*, 67–8. If T is the periodic time, and $a = \dfrac{v^2}{r} = \dfrac{4\pi^2 r}{T^2}$, the distance traversed (from rest) in rectilinear motion, in time T with acceleration a, is $\frac{1}{2}aT^2 = 2\pi^2 r$.

5. For if the planetary accelerations are as $\dfrac{r}{T^2}$ (above), and Kepler's Third Law states $T^2 = kr^3$, then the accelerations are as $\dfrac{1}{r^2}$.

6. R. T. Gunther, *Early Science in Oxford*, VI, 265–8 ; VII, 228, 27–8.

7. *Correspondence of Isaac Newton*, I, 284.

8. *Ibid.*, 416, 5 February 1676.

9. More's comments seem to miss the point, *Isaac Newton*, 177–9.

10. *Correspondence*, II, 309. Cf. Koyré, *Isis*, XLIII, 1952. This passage implies a derivation of the inverse-square law from Huygens' theorems on centrifugal force. Hooke probably reasoned that the peripheral velocity is proportional to the square root of the force (attraction), hence because the force is proportional to the inverse square of the distance the velocity is as the distance inversely. This is false, since $v = \sqrt{\dfrac{f}{d}}$, or, the velocity is as the square root of the distance inversely, which is what correctly follows from Kepler's Third Law.

11. *Ibid.*, 313, 17 January 1680.

12. The quotations are from the account of Newton's nephew-in-law John Conduitt (Ball, 26 etc.), which is confirmed by Halley's letters.

13. *Correspondence*, II, 437, Newton to Halley, 20 June 1686.

14. *Ibid.*, 435.

15. *Principia*, Motte-Cajori, 415–6.

16. *Ibid.*, (1687), 203. I have tried to improve upon Motte-Cajori, 202–3.

17. *Correspondence*, II, 438.

18. *Principia*, Preface.

19. *Opticks* (1931), 126–8, 154, 212.

20. *Principia*, Motte-Cajori, 397.

21. *Ibid.*, 230–1.

22. *Ibid.*, 543–4.

23. *Ibid.*, 545.

24. *Principia* (1687), 10. I give a paraphrase since a literal translation is almost unintelligible. Cf. Motte-Cajori, 11.

25. It might seem that this begs the question, since the sun's gravitational field is inferred from the centripetal forces of the planets' revolutions. The difficulty is removed, however, by considering the lunar motions, the effect of

the sun on the tides, etc., not to say Kepler's laws, which provide independent evidence for the solar gravitation.

CHAPTER XI: THE AGE OF NEWTON

H. G. Alexander, *The Leibniz-Clarke Correspondence* (Manchester, 1956); P. Brunet, *Les Physiciens Hollandais et la Méthode Expérimentale en France au 18ᵉ Siècle* (Paris, 1926); idem, *L'Introduction des Théories de Newton en France au 18ᵉ Siècle* (Paris, 1931); I. B. Cohen, *Franklin and Newton* (Philadelphia, 1956); A. R. Hall, " Correcting the *Principia* ", *Osiris*, XIII, 1958; C. Huygens, *Discours sur la Cause de la Pesanteur* (1690); H. Metzger, *Attraction universelle et Religion naturelle chez quelques commentateurs anglais de Newton* (Paris, 1938); idem, *Newton, Stahl, Boerhaave et la Doctrine Chimique* (Paris, 1930); P. Mouy, *op. cit.*, (Ch. IV); E. W. Strong, " Newtonian Explications of Natural Philosophy ", *Jour. Hist. Ideas*, XVII, 1957.

Notes (pp. 307–28)

1. Newton to Bentley, 25 February 1693; I. B. Cohen (ed.), *Papers and Letters*, 303.

2. No doubt because it is the only early portrait, and Newton "inclined to be fat in the latter part of his life ". Later artists portray Newton as a potentate, even a tyrant. But Kneller's face seems to agree well with the death-mask.

3. Possible exceptions are Christopher Wren (1680–2) and Robert Southwell (1690–5). At the time of their election both were better known for their other activities—in architecture and diplomacy—than for their youthful contributions to science.

4. Huygens, *Discours de la Pesanteur*, *Œuvres Complètes*, XXI, 475–6.

5. *Ibid.*, 474.

6. Fontenelle, *ed. cit.* (Ch. VI), 16 seq.

7. *Papers and Letters*, 457.

8. Brunet, *Introduction . . .* , 9.

9. From Kepler's Law, v^2r, not $\dfrac{v^2}{r}$ (the centrifugal force), is a constant.

10. This relation, though correct, is the product of some totally garbled mathematics. Instead of it he required, of course, $\dfrac{v^2}{r} = k$. *Recherche de la Vérité*, 1762, IV, 365–84. Brunet, *op. cit.*, 10–13; Mouy, 312. Neither of these writers mentions the crude errors of Villemot and Malebranche respectively.

11. Quoted by Mouy, 314.

12. *Papers and Letters*, 298, 302–3.

13. Alexander, 11.

14. *Ibid.*, 53.

15. *From the Closed World . . .* , Chapter XI.

16. Brunet, *Physiciens Hollandais*, 50.

17. Quoted by E. N. da C. Andrade, *Isaac Newton* (New York, 1950), 107.

18. Messrs. Sotheby, *Catalogue of the Newton Papers sold by Order of the Viscount Lymington* (London, 1936), Lot 210, p. 53.

19. *Ibid.*, 59 ; More, 247, 249.

20. He made hundred of thousands of words of transcript from such books ; cf. the Sotheby *Catalogue*.

21. Boas and Hall, *Arch. Int. d'Hist. des Sciences*, XI, 1958, 122–4.

22. Dark hints that Newton may have had insight into esoteric mysteries, that he had direct inspiration from the Unknown, that his wisdom was more than human and extended to knowledge of the structure of the atom and the dread possibility of nuclear force, are fruits of febrile imagination. Cf. Andrade, 105–8. There has been much exaggerated writing of Newton as a "mystic", for which there is no real evidence at all. When Newton spoke of the dangers of alchemy, for instance, he was aware like all thinking men of the economic and social consequences if gold should cease to be the measure of wealth and privilege.

23. *Opticks*, 394.

24. John Keill, "The Laws of Attraction", *Phil. Trans.*, no. 315, 1708.

25. The edition of 1709 was in Latin ; I give the title of the English version, 1712. The book provoked yet another dispute with the Cartesians over "esoteric principles".

26. J. Freind, *Chymical Lectures* (1712), 82–4.

27. Neither Keill nor Freind receive more than cursory allusions in I. B. Cohen's book on "experimental Newtonianism", *Franklin and Newton*.

28. *Opticks*, 381.

29. *Ibid.*, 387–8.

30. Peter Shaw, *New Method of Chemistry* (London, 1741), I, 173.

EPILOGUE : *Notes (pp. 329–43)*

1. Freind, *op. cit.*, 46.

2. John Harris, *Lexicon Technicum*, (London, 1704), s.v. "Engine". Freind was not listed among the subscribers to this dictionary, but "Mr Isaac Newton, Master of the Mint", was. The book is full of flattering references to him.

3. The reaction steam-turbine Hero actually invented was, of course, a different device.

4. E. Strauss, *Sir William Petty* (London, 1954), 114–19. Yet Oldenburg

could write to Petty that the Royal Society thought " themselves much concerned in that honor and reputation wch must attend an Invention of such publique & Generall benefit, by one of their Number ".

5. F. N. L. Poynter and W. J. Bishop, *A Seventeenth-Century Doctor and his Patients : John Symcotts, 1592?–1662.* Pub. Bedfordshire Hist. Record Soc., XXXI, 1951.

6. Sir W. Temple, *An Essay upon the Ancient and Modern Learning, Works* (1814), III, 476. Cf. R. F. Jones, *Ancients and Moderns* (St. Louis, 1936).

INDEX

INDEX

A CATALOG OF SELECTED
DOVER BOOKS
IN ALL FIELDS OF INTEREST

A CATALOG OF SELECTED DOVER
BOOKS IN ALL FIELDS OF INTEREST

LASERS AND HOLOGRAPHY, Winston E. Kock. Sound introduction to burgeoning field, expanded (1981) for second edition. 84 illustrations. 160pp. 5⅜ × 8¼. (EUK)   24041-X Pa. $3.50

FLORAL STAINED GLASS PATTERN BOOK, Ed Sibbett, Jr. 96 exquisite floral patterns—irises, poppie, lilies, tulips, geometrics, abstracts, etc.—adaptable to innumerable stained glass projects. 64pp. 8¼ × 11.   24259-5 Pa. $3.50

THE HISTORY OF THE LEWIS AND CLARK EXPEDITION, Meriwether Lewis and William Clark. Edited by Eliott Coues. Great classic edition of Lewis and Clark's day-by-day journals. Complete 1893 edition, edited by Eliott Coues from Biddle's authorized 1814 history. 1508pp. 5⅜ × 8½.
21268-8, 21269-6, 21270-X Pa. Three-vol. set $22.50

ORLEY FARM, Anthony Trollope. Three-dimensional tale of great criminal case. Original Millais illustrations illuminate marvelous panorama of Victorian society. Plot was author's favorite. 736pp. 5⅜ × 8½.   24181-5 Pa. $10.95

THE CLAVERINGS, Anthony Trollope. Major novel, chronicling aspects of British Victorian society, personalities. 16 plates by M. Edwards; first reprint of full text. 412pp. 5⅜ × 8½.   23464-9 Pa. $6.00

EINSTEIN'S THEORY OF RELATIVITY, Max Born. Finest semi-technical account; much explanation of ideas and math not readily available elsewhere on this level. 376pp. 5⅜ × 8½.   60769-0 Pa. $5.00

COMPUTABILITY AND UNSOLVABILITY, Martin Davis. Classic graduate-level introduction th theory of computability, usually referred to as theory of recurrent functions. New preface and appendix. 288pp. 5⅜ × 8½. 61471-9 Pa. $6.50

THE GODS OF THE EGYPTIANS, E.A. Wallis Budge. Never excelled for richness, fullness: all gods, goddesses, demons, mythical figures of Ancient Egypt; their legends, rites, incarnations, etc. Over 225 illustrations, plus 6 color plates. 988pp. 6⅛ × 9¼. (EBE)   22055-9, 22056-7 Pa., Two-vol. set $20.00

THE I CHING (THE BOOK OF CHANGES), translated by James Legge. Most penetrating divination manual ever prepared. Indispensable to study of early Oriental civilizations, to modern inquiring reader. 448pp. 5⅜ × 8½.
21062-6 Pa. $6.50

THE CRAFTSMAN'S HANDBOOK, Cennino Cennini. 15th-century handbook, school of Giotto, explains applying gold, silver leaf; gesso; fresco painting, grinding pigments, etc. 142pp. 6⅛ × 9¼.   20054-X Pa. $3.50

AN ATLAS OF ANATOMY FOR ARTISTS, Fritz Schider. Finest text, working book. Full text, plus anatomical illustrations; plates by great artists showing anatomy. 593 illustrations. 192pp. 7⅛ × 10¼.   20241-0 Pa. $6.50

EASY-TO-MAKE STAINED GLASS LIGHTCATCHERS, Ed Sibbett, Jr. 67 designs for most enjoyable ornaments: fruits, birds, teddy bears, trumpet, etc. Full size templates. 64pp. 8¼ × 11.   24081-9 Pa. $3.95

TRIAD OPTICAL ILLUSIONS AND HOW TO DESIGN THEM, Harry Turner. Triad explained in 32 pages of text, with 32 pages of Escher-like patterns on coloring stock. 92 figures. 32 plates. 64pp. 8¼ × 11.   23549-1 Pa. $2.95

SMOCKING: TECHNIQUE, PROJECTS, AND DESIGNS, Dianne Durand. Foremost smocking designer provides complete instructions on how to smock. Over 10 projects, over 100 illustrations. 56pp. 8¼ × 11.   23788-5 Pa. $2.00

AUDUBON'S BIRDS IN COLOR FOR DECOUPAGE, edited by Eleanor H. Rawlings. 24 sheets, 37 most decorative birds, full color, on one side of paper. Instructions, including work under glass. 56pp. 8¼ × 11.   23492-4 Pa. $3.95

THE COMPLETE BOOK OF SILK SCREEN PRINTING PRODUCTION, J.I. Biegeleisen. For commercial user, teacher in advanced classes, serious hobbyist. Most modern techniques, materials, equipment for optimal results. 124 illustrations. 253pp. 5⅜ × 8½.   21100-2 Pa. $4.50

A TREASURY OF ART NOUVEAU DESIGN AND ORNAMENT, edited by Carol Belanger Grafton. 577 designs for the practicing artist. Full-page, spots, borders, bookplates by Klimt, Bradley, others. 144pp. 8⅜ × 11¼.   24001-0 Pa. $5.95

ART NOUVEAU TYPOGRAPHIC ORNAMENTS, Dan X. Solo. Over 800 Art Nouveau florals, swirls, women, animals, borders, scrolls, wreaths, spots and dingbats, copyright-free. 100pp. 8⅛ × 11.   24366-4 Pa. $4.00

HAND SHADOWS TO BE THROWN UPON THE WALL, Henry Bursill. Wonderful Victorian novelty tells how to make flying birds, dog, goose, deer, and 14 others, each explained by a full-page illustration. 32pp. 6½ × 9¼. 21779-5 Pa. $1.50

AUDUBON'S BIRDS OF AMERICA COLORING BOOK, John James Audubon. Rendered for coloring by Paul Kennedy. 46 of Audubon's noted illustrations: red-winged black-bird, cardinal, etc. Original plates reproduced in full-color on the covers. Captions. 48pp. 8¼ × 11.   23049-X Pa. $2.25

SILK SCREEN TECHNIQUES, J.I. Biegeleisen, M.A. Cohn. Clear, practical, modern, economical. Minimal equipment (self-built), materials, easy methods. For amateur, hobbyist, 1st book. 141 illustrations. 185pp. 6⅛ × 9¼.   20433-2 Pa. $3.95

101 PATCHWORK PATTERNS, Ruby S. McKim. 101 beautiful, immediately useable patterns, full-size, modern and traditional. Also general information, estimating, quilt lore. 140 illustrations. 124pp. 7⅞ × 10¾.   20773-0 Pa. $3.50

READY-TO-USE FLORAL DESIGNS, Ed Sibbett, Jr. Over 100 floral designs (most in three sizes) of popular individual blossoms as well as bouquets, sprays, garlands. 64pp. 8¼ × 11.   23976-4 Pa. $2.95

AMERICAN WILD FLOWERS COLORING BOOK, Paul Kennedy. Planned coverage of 46 most important wildflowers, from Rickett's collection; instructive as well as entertaining. Color versions on covers. Captions. 48pp. 8¼ × 11.
20095-7 Pa. $2.50

CARVING DUCK DECOYS, Harry V. Shourds and Anthony Hillman. Detailed instructions and full-size templates for constructing 16 beautiful, marvelously practical decoys according to time-honored South Jersey method. 70pp. 9¼ × 12¼.
24083-5 Pa. $4.95

TRADITIONAL PATCHWORK PATTERNS, Carol Belanger Grafton. Cardboard cut-out pieces for use as templates to make 12 quilts: Buttercup, Ribbon Border, Tree of Paradise, nine more. Full instructions. 57pp. 8¼ × 11.
23015-5 Pa. $3.50

25 KITES THAT FLY, Leslie Hunt. Full, easy-to-follow instructions for kites made from inexpensive materials. Many novelties. 70 illustrations. 110pp. 5⅜ × 8½.
22550-X Pa. $2.25

PIANO TUNING, J. Cree Fischer. Clearest, best book for beginner, amateur. Simple repairs, raising dropped notes, tuning by easy method of flattened fifths. No previous skills needed. 4 illustrations. 201pp. 5⅜ × 8½.
23267-0 Pa. $3.50

EARLY AMERICAN IRON-ON TRANSFER PATTERNS, edited by Rita Weiss. 75 designs, borders, alphabets, from traditional American sources. 48pp. 8¼ × 11.
23162-3 Pa. $1.95

CROCHETING EDGINGS, edited by Rita Weiss. Over 100 of the best designs for these lovely trims for a host of household items. Complete instructions, illustrations. 48pp. 8¼ × 11.
24031-2 Pa. $2.25

FINGER PLAYS FOR NURSERY AND KINDERGARTEN, Emilie Poulsson. 18 finger plays with music (voice and piano); entertaining, instructive. Counting, nature lore, etc. Victorian classic. 53 illustrations. 80pp. 6½ × 9¼. 22588-7 Pa. $1.95

BOSTON THEN AND NOW, Peter Vanderwarker. Here in 59 side-by-side views are photographic documentations of the city's past and present. 119 photographs. Full captions. 122pp. 8¼ × 11.
24312-5 Pa. $6.95

CROCHETING BEDSPREADS, edited by Rita Weiss. 22 patterns, originally published in three instruction books 1939-41. 39 photos, 8 charts. Instructions. 48pp. 8¼ × 11.
23610-2 Pa. $2.00

HAWTHORNE ON PAINTING, Charles W. Hawthorne. Collected from notes taken by students at famous Cape Cod School; hundreds of direct, personal *apercus*, ideas, suggestions. 91pp. 5⅜ × 8½.
20653-X Pa. $2.50

THERMODYNAMICS, Enrico Fermi. A classic of modern science. Clear, organized treatment of systems, first and second laws, entropy, thermodynamic potentials, etc. Calculus required. 160pp. 5⅜ × 8½.
60361-X Pa. $4.00

TEN BOOKS ON ARCHITECTURE, Vitruvius. The most important book ever written on architecture. Early Roman aesthetics, technology, classical orders, site selection, all other aspects. Morgan translation. 331pp. 5⅜ × 8½. 20645-9 Pa. $5.50

THE CORNELL BREAD BOOK, Clive M. McCay and Jeanette B. McCay. Famed high-protein recipe incorporated into breads, rolls, buns, coffee cakes, pizza, pie crusts, more. Nearly 50 illustrations. 48pp. 8¼ × 11.
23995-0 Pa. $2.00

THE CRAFTSMAN'S HANDBOOK, Cennino Cennini. 15th-century handbook, school of Giotto, explains applying gold, silver leaf; gesso; fresco painting, grinding pigments, etc. 142pp. 6⅛ × 9¼.
20054-X Pa. $3.50

FRANK LLOYD WRIGHT'S FALLINGWATER, Donald Hoffmann. Full story of Wright's masterwork at Bear Run, Pa. 100 photographs of site, construction, and details of completed structure. 112pp. 9¼ × 10.
23671-4 Pa. $6.95

OVAL STAINED GLASS PATTERN BOOK, C. Eaton. 60 new designs framed in shape of an oval. Greater complexity, challenge with sinuous cats, birds, mandalas framed in antique shape. 64pp. 8¼ × 11.
24519-5 Pa. $3.50

CHILDREN'S BOOKPLATES AND LABELS, Ed Sibbett, Jr. 6 each of 12 types based on *Wizard of Oz, Alice,* nursery rhymes, fairy tales. Perforated; full color. 24pp. 8¼ × 11. 23538-6 Pa. $3.50

READY-TO-USE VICTORIAN COLOR STICKERS: 96 Pressure-Sensitive Seals, Carol Belanger Grafton. Drawn from authentic period sources. Motifs include heads of men, women, children, plus florals, animals, birds, more. Will adhere to any clean surface. 8pp. 8½ × 11. 24551-9 Pa. $2.95

CUT AND FOLD PAPER SPACESHIPS THAT FLY, Michael Grater. 16 colorful, easy-to-build spaceships that really fly. Star Shuttle, Lunar Freighter, Star Probe, 13 others. 32pp. 8¼ × 11. 23978-0 Pa. $2.50

CUT AND ASSEMBLE PAPER AIRPLANES THAT FLY, Arthur Baker. 8 aerodynamically sound, ready-to-build paper airplanes, designed with latest techniques. Fly *Pegasus, Daedalus, Songbird,* 5 other aircraft. Instructions. 32pp. 9¼ × 11¼. 24302-8 Pa. $3.95

SIDELIGHTS ON RELATIVITY, Albert Einstein. Two lectures delivered in 1920-21: *Ether and Relativity* and *Geometry and Experience.* Elegant ideas in non-mathematical form. 56pp. 5⅜ × 8½. 24511-X Pa. $2.25

FADS AND FALLACIES IN THE NAME OF SCIENCE, Martin Gardner. Fair, witty appraisal of cranks and quacks of science: Velikovsky, orgone energy, Bridey Murphy, medical fads, etc. 373pp. 5⅜ × 8½. 20394-8 Pa. $5.95

VACATION HOMES AND CABINS, U.S. Dept. of Agriculture. Complete plans for 16 cabins, vacation homes and other shelters. 105pp. 9 × 12. 23631-5 Pa. $4.95

HOW TO BUILD A WOOD-FRAME HOUSE, L.O. Anderson. Placement, foundations, framing, sheathing, roof, insulation, plaster, finishing—almost everything else. 179 illustrations. 223pp. 7⅞ × 10¾. 22954-8 Pa. $5.50

THE MYSTERY OF A HANSOM CAB, Fergus W. Hume. Bizarre murder in a hansom cab leads to engrossing investigation. Memorable characters, rich atmosphere. 19th-century bestseller, still enjoyable, exciting. 256pp. 5⅜ × 8. 21956-9 Pa. $4.00

MANUAL OF TRADITIONAL WOOD CARVING, edited by Paul N. Hasluck. Possibly the best book in English on the craft of wood carving. Practical instructions, along with 1,146 working drawings and photographic illustrations. 576pp. 6½ × 9¼. 23489-4 Pa. $8.95

WHITTLING AND WOODCARVING, E.J Tangerman. Best book on market; clear, full. If you can cut a potato, you can carve toys, puzzles, chains, etc. Over 464 illustrations. 293pp. 5⅜ × 8½. 20965-2 Pa. $4.95

AMERICAN TRADEMARK DESIGNS, Barbara Baer Capitman. 732 marks, logos and corporate-identity symbols. Categories include entertainment, heavy industry, food and beverage. All black-and-white in standard forms. 160pp. 8⅛ × 11. 23259-X Pa. $6.95

DECORATIVE FRAMES AND BORDERS, edited by Edmund V. Gillon, Jr. Largest collection of borders and frames ever compiled for use of artists and designers. Renaissance, neo-Greek, Art Nouveau, Art Deco, to mention only a few styles. 396 illustrations. 192pp. 8⅜ × 11¼. 22928-9 Pa. $6.00

THE MURDER BOOK OF J.G. REEDER, Edgar Wallace. Eight suspenseful stories by bestselling mystery writer of 20s and 30s. Features the donnish Mr. J.G. Reeder of Public Prosecutor's Office. 128pp. 5⅜ × 8½. (Available in U.S. only)
24374-5 Pa. $3.50

ANNE ORR'S CHARTED DESIGNS, Anne Orr. Best designs by premier needlework designer, all on charts: flowers, borders, birds, children, alphabets, etc. Over 100 charts, 10 in color. Total of 40pp. 8¼ × 11. 23704-4 Pa. $2.50

BASIC CONSTRUCTION TECHNIQUES FOR HOUSES AND SMALL BUILDINGS SIMPLY EXPLAINED, U.S. Bureau of Naval Personnel. Grading, masonry, woodworking, floor and wall framing, roof framing, plastering, tile setting, much more. Over 675 illustrations. 568pp. 6½ × 9¼. 20242-9 Pa. $8.95

MATISSE LINE DRAWINGS AND PRINTS, Henri Matisse. Representative collection of female nudes, faces, still lifes, experimental works, etc., from 1898 to 1948. 50 illustrations. 48pp. 8⅜ × 11¼. 23877-6 Pa. $2.50

HOW TO PLAY THE CHESS OPENINGS, Eugene Znosko-Borovsky. Clear, profound examinations of just what each opening is intended to do and how opponent can counter. Many sample games. 147pp. 5⅜ × 8½. 22795-2 Pa. $2.95

DUPLICATE BRIDGE, Alfred Sheinwold. Clear, thorough, easily followed account: rules, etiquette, scoring, strategy, bidding; Goren's point-count system, Blackwood and Gerber conventions, etc. 158pp. 5⅜ × 8½. 22741-3 Pa. $3.00

SARGENT PORTRAIT DRAWINGS, J.S. Sargent. Collection of 42 portraits reveals technical skill and intuitive eye of noted American portrait painter, John Singer Sargent. 48pp. 8¼ × 11⅛. 24524-1 Pa. $2.95

ENTERTAINING SCIENCE EXPERIMENTS WITH EVERYDAY OBJECTS, Martin Gardner. Over 100 experiments for youngsters. Will amuse, astonish, teach, and entertain. Over 100 illustrations. 127pp. 5⅜ × 8½. 24201-3 Pa. $2.50

TEDDY BEAR PAPER DOLLS IN FULL COLOR: A Family of Four Bears and Their Costumes, Crystal Collins. A family of four Teddy Bear paper dolls and nearly 60 cut-out costumes. Full color, printed one side only. 32pp. 9¼ × 12¼.
24550-0 Pa. $3.50

NEW CALLIGRAPHIC ORNAMENTS AND FLOURISHES, Arthur Baker. Unusual, multi-useable material: arrows, pointing hands, brackets and frames, ovals, swirls, birds, etc. Nearly 700 illustrations. 80pp. 8⅜ × 11¼.
24095-9 Pa. $3.75

DINOSAUR DIORAMAS TO CUT & ASSEMBLE, M. Kalmenoff. Two complete three-dimensional scenes in full color, with 31 cut-out animals and plants. Excellent educational toy for youngsters. Instructions; 2 assembly diagrams. 32pp. 9¼ × 12¼. 24541-1 Pa. $4.50

SILHOUETTES: A PICTORIAL ARCHIVE OF VARIED ILLUSTRATIONS, edited by Carol Belanger Grafton. Over 600 silhouettes from the 18th to 20th centuries. Profiles and full figures of men, women, children, birds, animals, groups and scenes, nature, ships, an alphabet. 144pp. 8⅜ × 11¼. 23781-8 Pa. $4.95

SURREAL STICKERS AND UNREAL STAMPS, William Rowe. 224 haunting, hilarious stamps on gummed, perforated stock, with images of elephants, geisha girls, George Washington, etc. 16pp. one side. 8¼ × 11.   24371-0 Pa. $3.50

GOURMET KITCHEN LABELS, Ed Sibbett, Jr. 112 full-color labels (4 copies each of 28 designs). Fruit, bread, other culinary motifs. Gummed and perforated. 16pp. 8¼ × 11.   24087-8 Pa. $2.95

PATTERNS AND INSTRUCTIONS FOR CARVING AUTHENTIC BIRDS, H.D. Green. Detailed instructions, 27 diagrams, 85 photographs for carving 15 species of birds so life-like, they'll seem ready to fly! 8¼ × 11.   24222-6 Pa. $2.75

FLATLAND, E.A. Abbott. Science-fiction classic explores life of 2-D being in 3-D world. 16 illustrations. 103pp. 5⅜ × 8.   20001-9 Pa. $2.00

DRIED FLOWERS, Sarah Whitlock and Martha Rankin. Concise, clear, practical guide to dehydration, glycerinizing, pressing plant material, and more. Covers use of silica gel. 12 drawings. 32pp. 5⅜ × 8½.   21802-3 Pa. $1.00

EASY-TO-MAKE CANDLES, Gary V. Guy. Learn how easy it is to make all kinds of decorative candles. Step-by-step instructions. 82 illustrations. 48pp. 8¼ × 11.

23881-4 Pa. $2.50

SUPER STICKERS FOR KIDS, Carolyn Bracken. 128 gummed and perforated full-color stickers: GIRL WANTED, KEEP OUT, BORED OF EDUCATION, X-RATED, COMBAT ZONE, many others. 16pp. 8¼ × 11.   24092-4 Pa. $2.50

CUT AND COLOR PAPER MASKS, Michael Grater. Clowns, animals, funny faces...simply color them in, cut them out, and put them together, and you have 9 paper masks to play with and enjoy. 32pp. 8¼ × 11.   23171-2 Pa. $2.25

A CHRISTMAS CAROL: THE ORIGINAL MANUSCRIPT, Charles Dickens. Clear facsimile of Dickens manuscript, on facing pages with final printed text. 8 illustrations by John Leech, 4 in color on covers. 144pp. 8⅜ × 11¼.

20980-6 Pa. $5.95

CARVING SHOREBIRDS, Harry V. Shourds & Anthony Hillman. 16 full-size patterns (all double-page spreads) for 19 North American shorebirds with step-by-step instructions. 72pp. 9¼ × 12¼.   24287-0 Pa. $4.95

THE GENTLE ART OF MATHEMATICS, Dan Pedoe. Mathematical games, probability, the question of infinity, topology, how the laws of algebra work, problems of irrational numbers, and more. 42 figures. 143pp. 5⅜ × 8½. (EBE)

22949-1 Pa. $3.50

READY-TO-USE DOLLHOUSE WALLPAPER, Katzenbach & Warren, Inc. Stripe, 2 floral stripes, 2 allover florals, polka dot; all in full color. 4 sheets (350 sq. in.) of each, enough for average room. 48pp. 8¼ × 11.   23495-9 Pa. $2.95

MINIATURE IRON-ON TRANSFER PATTERNS FOR DOLLHOUSES, DOLLS, AND SMALL PROJECTS, Rita Weiss and Frank Fontana. Over 100 miniature patterns: rugs, bedspreads, quilts, chair seats, etc. In standard dollhouse size. 48pp. 8¼ × 11.   23741-9 Pa. $1.95

THE DINOSAUR COLORING BOOK, Anthony Rao. 45 renderings of dinosaurs, fossil birds, turtles, other creatures of Mesozoic Era. Scientifically accurate. Captions. 48pp. 8¼ × 11.   24022-3 Pa. $2.50

THE BOOK OF WOOD CARVING, Charles Marshall Sayers. Still finest book for beginning student. Fundamentals, technique; gives 34 designs, over 34 projects for panels, bookends, mirrors, etc. 33 photos. 118pp. 7¾ × 10⅝. 23654-4 Pa. $3.95

CARVING COUNTRY CHARACTERS, Bill Higginbotham. Expert advice for beginning, advanced carvers on materials, techniques for creating 18 projects—mirthful panorama of American characters. 105 illustrations. 80pp. 8⅜ × 11.
24135-1 Pa. $2.50

300 ART NOUVEAU DESIGNS AND MOTIFS IN FULL COLOR, C.B. Grafton. 44 full-page plates display swirling lines and muted colors typical of Art Nouveau. Borders, frames, panels, cartouches, dingbats, etc. 48pp. 9⅜ × 12¼.
24354-0 Pa. $6.95

SELF-WORKING CARD TRICKS, Karl Fulves. Editor of *Pallbearer* offers 72 tricks that work automatically through nature of card deck. No sleight of hand needed. Often spectacular. 42 illustrations. 113pp. 5⅜ × 8½. 23334-0 Pa. $3.50

CUT AND ASSEMBLE A WESTERN FRONTIER TOWN, Edmund V. Gillon, Jr. Ten authentic full-color buildings on heavy cardboard stock in H-O scale. Sheriff's Office and Jail, Saloon, Wells Fargo, Opera House, others. 48pp. 9¼ × 12¼.
23736-2 Pa. $3.95

CUT AND ASSEMBLE AN EARLY NEW ENGLAND VILLAGE, Edmund V. Gillon, Jr. Printed in full color on heavy cardboard stock. 12 authentic buildings in H-O scale: Adams home in Quincy, Mass., Oliver Wight house in Sturbridge, smithy, store, church, others. 48pp. 9¼ × 12¼. 23536-X Pa. $4.95

THE TALE OF TWO BAD MICE, Beatrix Potter. Tom Thumb and Hunca Munca squeeze out of their hole and go exploring. 27 full-color Potter illustrations. 59pp. 4¼ × 5½. (Available in U.S. only) 23065-1 Pa. $1.75

CARVING FIGURE CARICATURES IN THE OZARK STYLE, Harold L. Enlow. Instructions and illustrations for ten delightful projects, plus general carving instructions. 22 drawings and 47 photographs altogether. 39pp. 8⅜ × 11.
23151-8 Pa. $2.50

A TREASURY OF FLOWER DESIGNS FOR ARTISTS, EMBROIDERERS AND CRAFTSMEN, Susan Gaber. 100 garden favorites lushly rendered by artist for artists, craftsmen, needleworkers. Many form frames, borders. 80pp. 8¼ × 11.
24096-7 Pa. $3.50

CUT & ASSEMBLE A TOY THEATER/THE NUTCRACKER BALLET, Tom Tierney. Model of a complete, full-color production of Tchaikovsky's classic. 6 backdrops, dozens of characters, familiar dance sequences. 32pp. 9⅜ × 12¼.
24194-7 Pa. $4.50

ANIMALS: 1,419 COPYRIGHT-FREE ILLUSTRATIONS OF MAMMALS, BIRDS, FISH, INSECTS, ETC., edited by Jim Harter. Clear wood engravings present, in extremely lifelike poses, over 1,000 species of animals. 284pp. 9 × 12.
23766-4 Pa. $9.95

MORE HAND SHADOWS, Henry Bursill. For those at their 'finger ends,'' 16 more effects—Shakespeare, a hare, a squirrel, Mr. Punch, and twelve more—each explained by a full-page illustration. Considerable period charm. 30pp. 6½ × 9¼.
21384-6 Pa. $1.95

JAPANESE DESIGN MOTIFS, Matsuya Co. Mon, or heraldic designs. Over 4000 typical, beautiful designs: birds, animals, flowers, swords, fans, geometrics; all beautifully stylized. 213pp. 11⅜ × 8¼. 22874-6 Pa. $7.95

THE TALE OF BENJAMIN BUNNY, Beatrix Potter. Peter Rabbit's cousin coaxes him back into Mr. McGregor's garden for a whole new set of adventures. All 27 full-color illustrations. 59pp. 4¼ × 5½. (Available in U.S. only) 21102-9 Pa. $1.75

THE TALE OF PETER RABBIT AND OTHER FAVORITE STORIES BOXED SET, Beatrix Potter. Seven of Beatrix Potter's best-loved tales including Peter Rabbit in a specially designed, durable boxed set. 4¼ × 5½. Total of 447pp. 158 color illustrations. (Available in U.S. only) 23903-9 Pa. $10.80

PRACTICAL MENTAL MAGIC, Theodore Annemann. Nearly 200 astonishing feats of mental magic revealed in step-by-step detail. Complete advice on staging, patter, etc. Illustrated. 320pp. 5⅜ × 8½. 24426-1 Pa. $5.95

CELEBRATED CASES OF JUDGE DEE (DEE GOONG AN), translated by Robert Van Gulik. Authentic 18th-century Chinese detective novel; Dee and associates solve three interlocked cases. Led to van Gulik's own stories with same characters. Extensive introduction. 9 illustrations. 237pp. 5⅜ × 8½. 23337-5 Pa. $4.50

CUT & FOLD EXTRATERRESTRIAL INVADERS THAT FLY, M. Grater. Stage your own lilliputian space battles. By following the step-by-step instructions and explanatory diagrams you can launch 22 full-color fliers into space. 36pp. 8¼ × 11. 24478-4 Pa. $2.95

CUT & ASSEMBLE VICTORIAN HOUSES, Edmund V. Gillon, Jr. Printed in full color on heavy cardboard stock, 4 authentic Victorian houses in H-O scale: Italian-style Villa, Octagon, Second Empire, Stick Style. 48pp. 9¼ × 12¼. 23849-0 Pa. $3.95

BEST SCIENCE FICTION STORIES OF H.G. WELLS, H.G. Wells. Full novel *The Invisible Man,* plus 17 short stories: "The Crystal Egg," "Aepyornis Island," "The Strange Orchid," etc. 303pp. 5⅜ × 8½. (Available in U.S. only) 21531-8 Pa. $4.95

TRADEMARK DESIGNS OF THE WORLD, Yusaku Kamekura. A lavish collection of nearly 700 trademarks, the work of Wright, Loewy, Klee, Binder, hundreds of others. 160pp. 8¾ × 8. (Available in U.S. only) 24191-2 Pa. $5.95

THE ARTIST'S AND CRAFTSMAN'S GUIDE TO REDUCING, ENLARGING AND TRANSFERRING DESIGNS, Rita Weiss. Discover, reduce, enlarge, transfer designs from any objects to any craft project. 12pp. plus 16 sheets special graph paper. 8¼ × 11. 24142-4 Pa. $3.50

TREASURY OF JAPANESE DESIGNS AND MOTIFS FOR ARTISTS AND CRAFTSMEN, edited by Carol Belanger Grafton. Indispensable collection of 360 traditional Japanese designs and motifs redrawn in clean, crisp black-and-white, copyright-free illustrations. 96pp. 8¼ × 11. 24435-0 Pa. $3.95

CHANCERY CURSIVE STROKE BY STROKE, Arthur Baker. Instructions and illustrations for each stroke of each letter (upper and lower case) and numerals. 54 full-page plates. 64pp. 8¼ × 11. 24278-1 Pa. $2.50

THE ENJOYMENT AND USE OF COLOR, Walter Sargent. Color relationships, values, intensities; complementary colors, illumination, similar topics. Color in nature and art. 7 color plates, 29 illustrations. 274pp. 5⅜ × 8½. 20944-X Pa. $4.95

SCULPTURE PRINCIPLES AND PRACTICE, Louis Slobodkin. Step-by-step approach to clay, plaster, metals, stone; classical and modern. 253 drawings, photos. 255pp. 8⅛ × 11. 22960-2 Pa. $7.50

VICTORIAN FASHION PAPER DOLLS FROM HARPER'S BAZAR, 1867-1898, Theodore Menten. Four female dolls with 28 elegant high fashion costumes, printed in full color. 32pp. 9¼ × 12¼. 23453-3 Pa. $3.50

FLOPSY, MOPSY AND COTTONTAIL: A Little Book of Paper Dolls in Full Color, Susan LaBelle. Three dolls and 21 costumes (7 for each doll) show Peter Rabbit's siblings dressed for holidays, gardening, hiking, etc. Charming borders, captions. 48pp. 4¼ × 5½. 24376-1 Pa. $2.25

NATIONAL LEAGUE BASEBALL CARD CLASSICS, Bert Randolph Sugar. 83 big-leaguers from 1909-69 on facsimile cards. Hubbell, Dean, Spahn, Brock plus advertising, info, no duplications. Perforated, detachable. 16pp. 8¼ × 11. 24308-7 Pa. $2.95

THE LOGICAL APPROACH TO CHESS, Dr. Max Euwe, et al. First-rate text of comprehensive strategy, tactics, theory for the amateur. No gambits to memorize, just a clear, logical approach. 224pp. 5⅜ × 8½. 24353-2 Pa. $4.50

MAGICK IN THEORY AND PRACTICE, Aleister Crowley. The summation of the thought and practice of the century's most famous necromancer, long hard to find. Crowley's best book. 436pp. 5⅜ × 8½. (Available in U.S. only) 23295-6 Pa. $6.50

THE HAUNTED HOTEL, Wilkie Collins. Collins' last great tale; doom and destiny in a Venetian palace. Praised by T.S. Eliot. 127pp. 5⅜ × 8½. 24333-8 Pa. $3.00

ART DECO DISPLAY ALPHABETS, Dan X. Solo. Wide variety of bold yet elegant lettering in handsome Art Deco styles. 100 complete fonts, with numerals, punctuation, more. 104pp. 8⅛ × 11. 24372-9 Pa. $4.50

CALLIGRAPHIC ALPHABETS, Arthur Baker. Nearly 150 complete alphabets by outstanding contemporary. Stimulating ideas; useful source for unique effects. 154 plates. 157pp. 8⅜ × 11¼. 21045-6 Pa. $5.95

ARTHUR BAKER'S HISTORIC CALLIGRAPHIC ALPHABETS, Arthur Baker. From monumental capitals of first-century Rome to humanistic cursive of 16th century, 33 alphabets in fresh interpretations. 88 plates. 96pp. 9 × 12. 24054-1 Pa. $4.50

LETTIE LANE PAPER DOLLS, Sheila Young. Genteel turn-of-the-century family very popular then and now. 24 paper dolls. 16 plates in full color. 32pp. 9¼ × 12¼. 24089-4 Pa. $3.50

TWENTY-FOUR ART NOUVEAU POSTCARDS IN FULL COLOR FROM CLASSIC POSTERS, Hayward and Blanche Cirker. Ready-to-mail postcards reproduced from rare set of poster art. Works by Toulouse-Lautrec, Parrish, Steinlen, Mucha, Cheret, others. 12pp. 8¼× 11. 24389-3 Pa. $2.95

READY-TO-USE ART NOUVEAU BOOKMARKS IN FULL COLOR, Carol Belanger Grafton. 30 elegant bookmarks featuring graceful, flowing lines, foliate motifs, sensuous women characteristic of Art Nouveau. Perforated for easy detaching. 16pp. 8¼ × 11. 24305-2 Pa. $2.95

FRUIT KEY AND TWIG KEY TO TREES AND SHRUBS, William M. Harlow. Fruit key covers 120 deciduous and evergreen species; twig key covers 160 deciduous species. Easily used. Over 300 photographs. 126pp. 5⅜ × 8½. 20511-8 Pa. $2.25

LEONARDO DRAWINGS, Leonardo da Vinci. Plants, landscapes, human face and figure, etc., plus studies for Sforza monument, *Last Supper*, more. 60 illustrations. 64pp. 8¼ × 11⅛. 23951-9 Pa. $2.75

CLASSIC BASEBALL CARDS, edited by Bert R. Sugar. 98 classic cards on heavy stock, full color, perforated for detaching. Ruth, Cobb, Durocher, DiMaggio, H. Wagner, 99 others. Rare originals cost hundreds. 16pp. 8¼ × 11. 23498-3 Pa. $3.25

TREES OF THE EASTERN AND CENTRAL UNITED STATES AND CANADA, William M. Harlow. Best one-volume guide to 140 trees. Full descriptions, woodlore, range, etc. Over 600 illustrations. Handy size. 288pp. 4½ × 6⅜. 20395-6 Pa. $3.95

JUDY GARLAND PAPER DOLLS IN FULL COLOR, Tom Tierney. 3 Judy Garland paper dolls (teenager, grown-up, and mature woman) and 30 gorgeous costumes highlighting memorable career. Captions. 32pp. 9¼ × 12¼. 24404-0 Pa. $3.50

GREAT FASHION DESIGNS OF THE BELLE EPOQUE PAPER DOLLS IN FULL COLOR, Tom Tierney. Two dolls and 30 costumes meticulously rendered. Haute couture by Worth, Lanvin, Paquin, other greats late Victorian to WWI. 32pp. 9¼ × 12¼. 24425-3 Pa. $3.50

FASHION PAPER DOLLS FROM GODEY'S LADY'S BOOK, 1840-1854, Susan Johnston. In full color: 7 female fashion dolls with 50 costumes. Little girl's, bridal, riding, bathing, wedding, evening, everyday, etc. 32pp. 9¼ × 12¼. 23511-4 Pa. $3.95

THE BOOK OF THE SACRED MAGIC OF ABRAMELIN THE MAGE, translated by S. MacGregor Mathers. Medieval manuscript of ceremonial magic. Basic document in Aleister Crowley, Golden Dawn groups. 268pp. 5⅜ × 8½. 23211-5 Pa. $5.00

PETER RABBIT POSTCARDS IN FULL COLOR: 24 Ready-to-Mail Cards, Susan Whited LaBelle. Bunnies ice-skating, coloring Easter eggs, making valentines, many other charming scenes. 24 perforated full-color postcards, each measuring 4¼ × 6, on coated stock. 12pp. 9 × 12. 24617-5 Pa. $2.95

CELTIC HAND STROKE BY STROKE, A. Baker. Complete guide creating each letter of the alphabet in distinctive Celtic manner. Covers hand position, strokes, pens, inks, paper, more. Illustrated. 48pp. 8¼ × 11. 24336-2 Pa. $2.50

KEYBOARD WORKS FOR SOLO INSTRUMENTS, G.F. Handel. 35 neglected works from Handel's vast oeuvre, originally jotted down as improvisations. Includes Eight Great Suites, others. New sequence. 174pp. 9⅜ × 12¼.
24338-9 Pa. $7.50

AMERICAN LEAGUE BASEBALL CARD CLASSICS, Bert Randolph Sugar. 82 stars from 1900s to 60s on facsimile cards. Ruth, Cobb, Mantle, Williams, plus advertising, info, no duplications. Perforated, detachable. 16pp. 8¼ × 11.
24286-2 Pa. $2.95

A TREASURY OF CHARTED DESIGNS FOR NEEDLEWORKERS, Georgia Gorham and Jeanne Warth. 141 charted designs: owl, cat with yarn, tulips, piano, spinning wheel, covered bridge, Victorian house and many others. 48pp. 8¼ × 11.
23558-0 Pa. $1.95

DANISH FLORAL CHARTED DESIGNS, Gerda Bengtsson. Exquisite collection of over 40 different florals: anemone, Iceland poppy, wild fruit, pansies, many others. 45 illustrations. 48pp. 8¼ × 11.
23957-8 Pa. $1.75

OLD PHILADELPHIA IN EARLY PHOTOGRAPHS 1839-1914, Robert F. Looney. 215 photographs: panoramas, street scenes, landmarks, President-elect Lincoln's visit, 1876 Centennial Exposition, much more. 230pp. 8⅜ × 11¾.
23345-6 Pa. $9.95

PRELUDE TO MATHEMATICS, W.W. Sawyer. Noted mathematician's lively, stimulating account of non-Euclidean geometry, matrices, determinants, group theory, other topics. Emphasis on novel, striking aspects. 224pp. 5⅜ × 8½.
24401-6 Pa. $4.50

ADVENTURES WITH A MICROSCOPE, Richard Headstrom. 59 adventures with clothing fibers, protozoa, ferns and lichens, roots and leaves, much more. 142 illustrations. 232pp. 5⅜ × 8½.
23471-1 Pa. $3.95

IDENTIFYING ANIMAL TRACKS: MAMMALS, BIRDS, AND OTHER ANIMALS OF THE EASTERN UNITED STATES, Richard Headstrom. For hunters, naturalists, scouts, nature-lovers. Diagrams of tracks, tips on identification. 128pp. 5⅜ × 8.
24442-3 Pa. $3.50

VICTORIAN FASHIONS AND COSTUMES FROM HARPER'S BAZAR, 1867-1898, edited by Stella Blum. Day costumes, evening wear, sports clothes, shoes, hats, other accessories in over 1,000 detailed engravings. 320pp. 9⅜ × 12¼.
22990-4 Pa. $10.95

EVERYDAY FASHIONS OF THE TWENTIES AS PICTURED IN SEARS AND OTHER CATALOGS, edited by Stella Blum. Actual dress of the Roaring Twenties, with text by Stella Blum. Over 750 illustrations, captions. 156pp. 9 × 12.
24134-3 Pa. $8.50

HALL OF FAME BASEBALL CARDS, edited by Bert Randolph Sugar. Cy Young, Ted Williams, Lou Gehrig, and many other Hall of Fame greats on 92 full-color, detachable reprints of early baseball cards. No duplication of cards with *Classic Baseball Cards.* 16pp. 8¼ × 11.
23624-2 Pa. $3.50

THE ART OF HAND LETTERING, Helm Wotzkow. Course in hand lettering, Roman, Gothic, Italic, Block, Script. Tools, proportions, optical aspects, individual variation. Very quality conscious. Hundreds of specimens. 320pp. 5⅜ × 8½.
21797-3 Pa. $4.95

HOW THE OTHER HALF LIVES, Jacob A. Riis. Journalistic record of filth, degradation, upward drive in New York immigrant slums, shops, around 1900. New edition includes 100 original Riis photos, monuments of early photography. 233pp. 10 × 7⅞. 22012-5 Pa. $7.95

CHINA AND ITS PEOPLE IN EARLY PHOTOGRAPHS, John Thomson. In 200 black-and-white photographs of exceptional quality photographic pioneer Thomson captures the mountains, dwellings, monuments and people of 19th-century China. 272pp. 9⅜ × 12¼. 24393-1 Pa. $12.95

GODEY COSTUME PLATES IN COLOR FOR DECOUPAGE AND FRAMING, edited by Eleanor Hasbrouk Rawlings. 24 full-color engravings depicting 19th-century Parisian haute couture. Printed on one side only. 56pp. 8¼ × 11. 23879-2 Pa. $3.95

ART NOUVEAU STAINED GLASS PATTERN BOOK, Ed Sibbett, Jr. 104 projects using well-known themes of Art Nouveau: swirling forms, florals, peacocks, and sensuous women. 60pp. 8¼ × 11. 23577-7 Pa. $3.50

QUICK AND EASY PATCHWORK ON THE SEWING MACHINE: Susan Aylsworth Murwin and Suzzy Payne. Instructions, diagrams show exactly how to machine sew 12 quilts. 48pp. of templates. 50 figures. 80pp. 8¼ × 11. 23770-2 Pa. $3.50

THE STANDARD BOOK OF QUILT MAKING AND COLLECTING, Marguerite Ickis. Full information, full-sized patterns for making 46 traditional quilts, also 150 other patterns. 483 illustrations. 273pp. 6⅞ × 9⅝. 20582-7 Pa. $5.95

LETTERING AND ALPHABETS, J. Albert Cavanagh. 85 complete alphabets lettered in various styles; instructions for spacing, roughs, brushwork. 121pp. 8¾ × 8. 20053-1 Pa. $3.95

LETTER FORMS: 110 COMPLETE ALPHABETS, Frederick Lambert. 110 sets of capital letters; 16 lower case alphabets; 70 sets of numbers and other symbols. 110pp. 8⅛ × 11. 22872-X Pa. $4.50

ORCHIDS AS HOUSE PLANTS, Rebecca Tyson Northen. Grow cattleyas and many other kinds of orchids—in a window, in a case, or under artificial light. 63 illustrations. 148pp. 5⅜ × 8½. 23261-1 Pa. $2.95

THE MUSHROOM HANDBOOK, Louis C.C. Krieger. Still the best popular handbook. Full descriptions of 259 species, extremely thorough text, poisons, folklore, etc. 32 color plates; 126 other illustrations. 560pp. 5⅜ × 8½. 21861-9 Pa. $8.50

THE DORÉ BIBLE ILLUSTRATIONS, Gustave Doré. All wonderful, detailed plates: Adam and Eve, Flood, Babylon, life of Jesus, etc. Brief King James text with each plate. 241 plates. 241pp. 9 × 12. 23004-X Pa. $8.95

THE BOOK OF KELLS: Selected Plates in Full Color, edited by Blanche Cirker. 32 full-page plates from greatest manuscript-icon of early Middle Ages. Fantastic, mysterious. Publisher's Note. Captions. 32pp. 9¾ × 12¼. 24345-1 Pa. $4.50

THE PERFECT WAGNERITE, George Bernard Shaw. Brilliant criticism of the Ring Cycle, with provocative interpretation of politics, economic theories behind the Ring. 136pp. 5⅜ × 8½. (Available in U.S. only) 21707-8 Pa. $3.00

THE RIME OF THE ANCIENT MARINER, Gustave Doré, S.T. Coleridge. Doré's finest work, 34 plates capture moods, subtleties of poem. Full text. 77pp. 9¼ × 12. 22305-1 Pa. $4.95

SONGS OF INNOCENCE, William Blake. The first and most popular of Blake's famous "Illuminated Books," in a facsimile edition reproducing all 31 brightly colored plates. Additional printed text of each poem. 64pp. 5¼ × 7.
22764-2 Pa. $3.50

AN INTRODUCTION TO INFORMATION THEORY, J.R. Pierce. Second (1980) edition of most impressive non-technical account available. Encoding, entropy, noisy channel, related areas, etc. 320pp. 5⅜ × 8½. 24061-4 Pa. $4.95

THE DIVINE PROPORTION: A STUDY IN MATHEMATICAL BEAUTY, H.E. Huntley. "Divine proportion" or "golden ratio" in poetry, Pascal's triangle, philosophy, psychology, music, mathematical figures, etc. Excellent bridge between science and art. 58 figures. 185pp. 5⅜ × 8½. 22254-3 Pa. $3.95

THE DOVER NEW YORK WALKING GUIDE: From the Battery to Wall Street, Mary J. Shapiro. Superb inexpensive guide to historic buildings and locales in lower Manhattan: Trinity Church, Bowling Green, more. Complete Text; maps. 36 illustrations. 48pp. 3⅞ × 9¼. 24225-0 Pa. $2.50

NEW YORK THEN AND NOW, Edward B. Watson, Edmund V. Gillon, Jr. 83 important Manhattan sites: on facing pages early photographs (1875-1925) and 1976 photos by Gillon. 172 illustrations. 171pp. 9¼ × 10. 23361-8 Pa. $7.95

HISTORIC COSTUME IN PICTURES, Braun & Schneider. Over 1450 costumed figures from dawn of civilization to end of 19th century. English captions. 125 plates. 256pp. 8⅜ × 11¼. 23150-X Pa. $7.50

VICTORIAN AND EDWARDIAN FASHION: A Photographic Survey, Alison Gernsheim. First fashion history completely illustrated by contemporary photographs. Full text plus 235 photos, 1840-1914, in which many celebrities appear. 240pp. 6½ × 9¼. 24205-6 Pa. $6.00

CHARTED CHRISTMAS DESIGNS FOR COUNTED CROSS-STITCH AND OTHER NEEDLECRAFTS, Lindberg Press. Charted designs for 45 beautiful needlecraft projects with many yuletide and wintertime motifs. 48pp. 8¼ × 11. 24356-7 Pa. $2.50

101 FOLK DESIGNS FOR COUNTED CROSS-STITCH AND OTHER NEEDLE-CRAFTS, Carter Houck. 101 authentic charted folk designs in a wide array of lovely representations with many suggestions for effective use. 48pp. 8¼ × 11.
24369-9 Pa. $2.25

FIVE ACRES AND INDEPENDENCE, Maurice G. Kains. Great back-to-the-land classic explains basics of self-sufficient farming. The one book to get. 95 illustrations. 397pp. 5⅜ × 8½. 20974-1 Pa. $4.95

A MODERN HERBAL, Margaret Grieve. Much the fullest, most exact, most useful compilation of herbal material. Gigantic alphabetical encyclopedia, from aconite to zedoary, gives botanical information, medical properties, folklore, economic uses, and much else. Indispensable to serious reader. 161 illustrations. 888pp. 6½ × 9¼. (Available in U.S. only) 22798-7, 22799-5 Pa., Two-vol. set $16.45

SOURCE BOOK OF MEDICAL HISTORY, edited by Logan Clendening, M.D. Original accounts ranging from Ancient Egypt and Greece to discovery of X-rays: Galen, Pasteur, Lavoisier, Harvey, Parkinson, others. 685pp. 5⅜ × 8½.
20621-1 Pa. $10.95

THE ROSE AND THE KEY, J.S. Lefanu. Superb mystery novel from Irish master. Dark doings among an ancient and aristocratic English family. Well-drawn characters; capital suspense. Introduction by N. Donaldson. 448pp. 5⅜ × 8½.
24377-X Pa. $6.95

SOUTH WIND, Norman Douglas. Witty, elegant novel of ideas set on languorous Mediterranean island of Nepenthe. Elegant prose, glittering epigrams, mordant satire. 1917 masterpiece. 416pp. 5⅜ × 8½. (Available in U.S. only)
24361-3 Pa. $5.95

RUSSELL'S CIVIL WAR PHOTOGRAPHS, Capt. A.J. Russell. 116 rare Civil War Photos: Bull Run, Virginia campaigns, bridges, railroads, Richmond, Lincoln's funeral car. Many never seen before. Captions. 128pp. 9⅜ × 12¼.
24283-8 Pa. $6.95

PHOTOGRAPHS BY MAN RAY: 105 Works, 1920-1934. Nudes, still lifes, landscapes, women's faces, celebrity portraits (Dali, Matisse, Picasso, others), rayographs. Reprinted from rare gravure edition. 128pp. 9⅜ × 12¼. (Available in U.S. only)
23842-3 Pa. $7.95

STAR NAMES: THEIR LORE AND MEANING, Richard H. Allen. Star names, the zodiac, constellations: folklore and literature associated with heavens. The basic book of its field, fascinating reading. 563pp. 5⅜ × 8½.
21079-0 Pa. $7.95

BURNHAM'S CELESTIAL HANDBOOK, Robert Burnham, Jr. Thorough guide to the stars beyond our solar system. Exhaustive treatment. Alphabetical by constellation: Andromeda to Cetus in Vol. 1; Chamaeleon to Orion in Vol. 2; and Pavo to Vulpecula in Vol. 3. Hundreds of illustrations. Index in Vol. 3. 2000pp. 6⅛ × 9¼.
23567-X, 23568-8, 23673-0 Pa. Three-vol. set $36.85

THE ART NOUVEAU STYLE BOOK OF ALPHONSE MUCHA, Alphonse Mucha. All 72 plates from *Documents Decoratifs* in original color. Stunning, essential work of Art Nouveau. 80pp. 9⅜ × 12¼.
24044-4 Pa. $7.95

DESIGNS BY ERTE; FASHION DRAWINGS AND ILLUSTRATIONS FROM "HARPER'S BAZAR," Erte. 310 fabulous line drawings and 14 *Harper's Bazar* covers, 8 in full color. Erte's exotic temptresses with tassels, fur muffs, long trains, coifs, more. 129pp. 9⅜ × 12¼.
23397-9 Pa. $6.95

HISTORY OF STRENGTH OF MATERIALS, Stephen P. Timoshenko. Excellent historical survey of the strength of materials with many references to the theories of elasticity and structure. 245 figures. 452pp. 5⅜ × 8½.
61187-6 Pa. $8.95

Prices subject to change without notice.
Available at your book dealer or write for free catalog to Dept. GI, Dover Publications, Inc., 31 East 2nd St. Mineola, N.Y. 11501. Dover publishes more than 175 books each year on science, elementary and advanced mathematics, biology, music, art, literary history, social sciences and other areas.